Studies in Computational Intelligence 436

For further volumes:
http://www.springer.com/series/7092

Evangelos Karapanos

Modeling Users' Experiences with Interactive Systems

Author
Dr. Evangelos Karapanos
Madeira Interactive
Technologies Institute
Campus da Penteada
Funchal
Portugal

ISSN 1860-949X e-ISSN 1860-9503
ISBN 978-3-642-43360-3 ISBN 978-3-642-31000-3 (eBook)
DOI 10.1007/978-3-642-31000-3
Springer Heidelberg New York Dordrecht London

Printed on acid-free paper

Springer is part of Springer Science+Business Media (www.springer.com)

Foreword

Recent research in soft reliability has revealed that almost 50% of product returns in the consumer electronics market should not be attributed to a violation of product specifications, but rather to flawed design decisions, mostly made in the concept design phase. Current research in the field of Human-Computer Interaction recognizes that usability and usefulness are not the only relevant product attributes, and that more attention needs to be paid to other relevant issues of product quality, such as the aesthetic, emotional, and social aspects of a product. But how can we account for such non-instrumental quality aspects early in the design process? And how can we design products that comply with the values of a mixed audience? This book is not pretending to provide final answers to such questions, but rather explores the issues and leads the way to some interesting new challenges. While many authors have been quick to adopt (rather superficially) new hot topics such as user experience design, this book aims to show the reader which interesting and non-trivial issues are waiting to be explored.

The need for new experimental methods and analysis procedures that can provide insight in users attitudes towards early concepts and that can account for diversity in preferences, both across users and across time, was even hardly realized at the time when the research which is documented in this book project started. In this sense, some of the major contributions of the book are in helping to define the issues involved. Next to proposing specific methods, the author has contributed by bringing together information from very diverse disciplines such as: human-computer interaction, psychophysics & psychometrics, statistics, personal construct theory, memory and experience reconstruction, latent semantic analysis, etc. In this sense this book is a prime example of multidisciplinary research, illustrating an excellent knowledge of relevant literature, and demonstrating an ability to interpret and restructure such knowledge.

Of course the book doesnt only identify problems but also makes very valuable contributions for how to arrive at solutions. Chapter 2 introduces the concept of personal attribute judgments and a related experimental method called the repertory grid technique (RGT). While this method is well-known in some (more naturalistic oriented) areas of research, it is less so in mainstream (more positivistic oriented) areas of research. The specific contribution of chapter 2 is showing how the RGT can be combined with quantitative methods. This leads, amongst others, to new statistical challenges for finding perspectives that are intermediate between global averages (across all participants in an experiment) and individual judgments. In chapter 3, the proposed approach is applied in a concrete industrial case to illustrate how developers and intended users of a new system can indeed develop very different perspectives on what should be considered important issues. It is especially this difference in perspective that is held (at least partly) responsible for the above-cited soft reliability problem.

Chapter 4 makes a switch from the diversity between subjects to the change in user experience over time. While many researchers have underlined the importance of longitudinal studies, very few examples of such studies have actually been reported. The example case reported in chapter 4 illustrates a method for how to conduct such a (5 week) study but, more importantly, also provides a theoretical framework for how to analyze and structure the (mostly qualitative) observations that are collected in such a study. The outcomes of the study confirm our intuition that product experiences can indeed change substantially, especially in the first couple of weeks after a product purchases. While the method reported in chapter 4 is very valuable for in-depth scientific studies, commercial design practices are probably better served by large-scale surveys. How to assist users of products in recollecting and reporting experiences with their product in a way that is suitable for such surveys is the issue being addressed in chapter 5. Again, the chapter provides a combination of a method, implemented in the software tool iScale, and a well-structured scientific motivation, introducing the reader in an efficient way to relevant issues on memory and memory recollection. It is shown how different theories on memory recollection can lead to different variations of a method (more specifically, a constructive and a value-account approach), and how a scientific experiment can be used to make an informed choice between such possible alternatives.

The last chapter is clearly also the most speculative one. If the method proposed in chapter 5 can be introduced successfully, then the amount of reported user experiences can be expected to easily grow to a size where processing by a human analyst is simply no longer feasible. This raises the issue of how to extract design-relevant knowledge from such a large database of textual information. Chapter 6 discusses both a fully automated and a semi-automated approach for addressing this issue. Especially the semi-automated approach seems very promising, but substantial additional research will be required to settle the issue. While the problems are non-straightforward, the potential impact on design practices, and the related economical benefits, can hardly be overrated.

Im convinced that this book will provide food for thought for both novice and experienced researchers and designers interested in contributing to the new and growing area of user experiences. For me personally, I enjoyed very much coaching the author, Evangelos Karapanos in his explorations that are reported in this book.

Eindhoven, March 2012 Jean-Bernard Martens

Acknowledgements

The work described in this book has been carried out under the auspices of the J.F. Schouten Graduate School of User-System Interaction Research, and has been sponsored by the Dutch Ministry of Economic Affairs within the IOP-IPCR program, through the Soft Reliability project, under the auspices of Philips and Océ. Excerpts of this book have been previously published in Karapanos and Martens (2007) and Karapanos et al. (2009a, 2008a, 2009c,b).

Contents

Chapter 1
Introduction

48% of returned products are not attributed to a violation of product specifications (Den Ouden et al., 2006). This finding was the initial motivation for this research project. Brombacher, den Ouden and colleagues (e.g., Den Ouden et al., 2006; Brombacher et al., 2005; Koca et al., 2009) found that an alarmingly increasing number of returned products, in 2002 covering 48% of returned products, are technically fully functional, i.e. according to specifications, but they are returned on the basis of failing to satisfy users' true needs (28%), or purely because of users' remorse (20%) (Den Ouden et al., 2006). Brombacher et al. (2005) introduced the term 'Soft Reliability' to refer to these situations where *"in spite of meeting with the explicit product specifications, a customer explicitly complains on the (lack of) functionality of the product"*.

How is this finding different from Suchman's well-known case at XEROX in the '80s where users were found to have problems in using a feature-rich photocopier (c.f., Suchman, 2006)? While product designers were aiming at improving the instructions for using the product, Suchman argued that learning is an inherently problematic activity, and suggested that *"no matter how improved the machine interface or instruction set might be, this would never eliminate the need for active sense-making on the part of prospective users"* (Suchman, 2006, , p. 9). Since the '80s, a wealth of theories, methods, and design guidelines have been developed in the field of Human-Computer Interaction with the aim of making products more easy to learn and use in the long run. Thus, one might wonder, do Brombacher's and den Ouden's findings replicate what was found at XEROX almost 30 years ago, or do they introduce a new and as yet unaddressed problem? Should these consumer complaints be attributed to bad design practices, to apparently inescapable interaction flaws in first-time use, or do they suggest a new emerging problem in the user acceptance of interactive products?

Den Ouden et al. (2006) identified a number of trends in the consumer electronics (CE) industry that have resulted in radical differences of the current market in comparison to that in the '90s. They argued that the emphasis in the CE industry has shifted from the production of high volumes at competitive prices to the introduction of highly innovative products at higher prices. This leads to a shift in the

E. Karapanos: Modeling Users' Experiences with Interact. Syst., SCI 436, pp. 1–16.
springerlink.com

main uncertainty in new product development projects; while in the '90s the uncertainty related to the technology in relation to cost-effective mass production, in the current market the dominant uncertainty relates to the attractiveness of the product and users' expectations about the product functions.

These trends are reflected in the development of the field of Human-Computer Interaction, from the study of usability as a critical factor to the acceptance of interactive products, to a more holistic understanding of users' experiences with interactive products, leading to the study of new concepts like pleasure (Jordan, 2000), fun (Blythe et al., 2003), aesthetics (Tractinsky et al., 2000) and hedonic qualities (Hassenzahl, 2004). While a wealth of techniques and methods exist for ensuring the usability of interactive products, research on user experience evaluation methods is only at its infancy. This book aims at highlighting methodological issues in user experience evaluation and proposes a number of methods for inquiring into users' experiences with interactive products.

1.1 From Usability to Experience

The field of Human-Computer Interaction was for a long time identified as the field of usability engineering. Usability was seen as critical to user acceptance and a wealth of principles (e.g., Norman, 1988), design guidelines (e.g., Nielsen and Bellcore, 1992) and evaluation techniques (e.g., Dix et al., 2004) have become instrumental in the development of usable products. The field of usability engineering readily acknowledged the dual nature of the usability concept: its *objective* and *subjective* side. One of the most dominant definitions of usability, for instance, the ISO 9241-11 standard (1996) defines usability as

> *"the extend to which a product can be used by specific users to achieve specified goals with effectiveness, efficiency and satisfaction in a specified context of use".*

Effectiveness represents the accuracy and completeness with which users achieve certain goals and is typically measured through observed error rates, binary task completion, and quality of the outcome for a given task (see Frøkjær et al., 2000; Hornbæk and Law, 2007). Efficiency can be characterized as effectiveness in relation to resources spent and is typically measured through task completion time and learning time. One can note that both these components, effectiveness and efficiency represent the objective side of usability. The third component, user satisfaction, represents users' comfort in using and overall attitude to the product and is typically measured through psychometric scales on overall preference, product quality perceptions and specific attitudes towards the interface (see Hornbæk and Law, 2007).

An assumption underlying the distinction between the subjective and objective side of usability was that these two would strongly correlate. Nielsen and Levy (1994), in a meta-analysis of a selected set of 57 studies found that in 75% of the cases, users' overall preference was strongly related to overall performance. These findings have, however, been repeatedly questioned in subsequent studies suggesting that subjective perceptions of usability are generally not correlated with objective measures and seem to measure something else than merely effectiveness and efficiency (Frøkjær et al., 2000; Hornbæk and Law, 2007; Kissel, 1995; Bailey, 1993). This limited view on user satisfaction as a consequence of objective performance was criticized by Hassenzahl et al. (2000):

> *"...it seems as if satisfaction is conceived as a consequence of user experienced effectiveness and efficiency rather than a design goal in itself. This implies that assuring efficiency and effectiveness alone guarantees user satisfaction."*

Subsequently, a set of studies tried to explain these observed discrepancies between objective and subjective usability. Kurosu and Kashimura (1995) asked participants to rate several Automatic Teller Machine (ATM) designs on both functional and aesthetic aspects. They found apparent usability, i.e. subjective judgments on usability, to correlate more strongly with aesthetic judgments than with the systems' inherent usability, i.e. objective design parameters that were expected to affects users' performance in using the systems. Tractinsky (1997) and Tractinsky et al. (2000) replicated this study and found that these effects persisted both across different cultural backgrounds, as well as after participants had experienced the systems.

These early findings suggested that users' experiences with products go beyond the effectiveness and efficiency in product usage. Consequently, the field of Human-Computer Interaction quested for new *concepts*, *measures* and *methods* in capturing a more holistic view on user experience. This development has gone hand-in-hand with a shift in the contexts of study, from *professional* to *personal* (e.g., Jordan, 2000) and *social* (e.g., Forlizzi, 2007; Markopoulos et al., 2004), and in the design paradigm from *product* to *experience design* (e.g., Buxton, 2007; Zimmerman et al., 2007; Forlizzi et al., 2008).

1.2 Two Distinct Approaches in User Experience Research

User experience has become central to the design and evaluation of interactive products. It reflects a paradigm shift in the subject of product design and evaluation. Buxton (2007) argues the following:

> *"Ultimately, we are deluding ourselves if we think that the products that we design are the "things" that we sell, rather than the individual, social and cultural experience that they engender, and the value and impact that they have. Design that ignores this is not worthy of the name"*

However, user experience research is often criticized for at least two things: a) for the lack of a commonly agreed definition of the notion of experience, and b) for being identical, conceptually or methodologically, to traditional usability research. Indeed, Hassenzahl (2008) and Wright and Blythe (2007), some of the strong proponents of user experience research, criticize the use of the term *user experience* in cases where the focus still lies in traditional usability evaluation, thus reducing the richness of experience to behavioral logs and task-focused evaluations (Wright and Blythe, 2007). As Hassenzahl (2008) argues:

> *"While UX seems ubiquitous in industry, a closer look reveals that it is treated mainly as a synonym of usability and user-centered design"*

Conversely, as Hassenzahl (2008) argues, academics "emphasize the differences between traditional usability and user experience". A number of frameworks have tried to conceptualize how experiences are formed (e.g., Forlizzi and Ford, 2000; Wright and McCarthy, 2004; Norman, 2004; Hassenzahl, 2008) and tentative definitions of experience have been proposed (Forlizzi and Battarbee, 2004; Hassenzahl and Tractinsky, 2006; Hassenzahl, 2008).

Hassenzahl (2008) defines user experience as *"a momentary, primarily evaluative feeling (good-bad) while interacting with a product or service"* which is "a consequence of users' internal state (e.g. predispositions, expectations, needs, motivation, mood), the characteristics of the designed system (e.g. complexity, purpose, usability, functionality), and the context within which the interaction occurs (e.g. organizational/social setting, meaningfulness of the activity, voluntariness of use)" (Hassenzahl and Tractinsky, 2006). One may argue that such a definition, while being perhaps the best of what the field of user experience can offer at the moment, is far from being mature or useful for grounding measures, methods and principles in the design and evaluation of interactive products. Yet, a common ground has been established among various disciplines and schools of thoughts in the emerging field of user experience, perhaps due to "a history of use of the term in ordinary conversation and philosophy discourse" (Wright and Blythe, 2007). A number of researchers have tried to identify the dominant schools of thought and several classifications have been proposed (e.g. Battarbee and Koskinen, 2005; Hassenzahl and Tractinsky, 2006; Blythe et al., 2007).

We employ the distinction from Blythe et al. (2007) between *reductionist* approaches that have their roots in cognitive psychology, and *holistic* approaches that are grounded in pragmatist philosophy and phenomenology. As it will become apparent in section 1.4, we are primarily interested in distinct issues that these two approaches pose when one is concerned about methodology for understanding user experiences.

1.2.1 Reductionist Approaches

Reductionist approaches in user experience maintain a similar paradigm to *usability* (ISO, 1996) and *Technology Acceptance* research (see Venkatesh et al., 2003) in trying to identify distinct psychological constructs and propose and empirically test causal relations between them.

One of the first and well cited studies in user experience, grounded on reductionism, is that of Tractinsky (1997). Tractinsky was puzzled by the findings of Kurosu and Kashimura (1995) who suggested that subjective perceptions of usability relate more to the beauty of the product than to its actual, i.e. inherent, usability. This finding highlighted the importance of aesthetics in interactive products, an opinion that found Tractinsky resonant with. Tractinsky, however, predicted that this might not pertain over different cultures, taking into account that the study of Kurosu and Kashimura (1995) was conducted within the Japanese culture which is known for its aesthetic tradition. Tractinsky (1997) replicated the experiment using the same stimuli, but now in an Israeli context. His initial prediction was not confirmed as the findings of Kurosu and Kashimura (1995) were reconfirmed in this alternative setting. One possible criticism of both studies could be that the user judgments were elicited merely on the basis of the visual appearance of the interface without experiencing the systems. In a subsequent study, Tractinsky et al. (2000) elicited users' perceptions both before and after interacting with a computer simulation of an interface of an Automatic Teller Machine (ATM). The results suggested that the aesthetics of the interface also impacted the post-use perceptions of usability. Subsequent work has supported the dominance of beauty in users' preferences (e.g., Schenkman and Jönsson, 2000; Lindgaard and Dudek, 2003; Lindgaard et al., 2006; Tractinsky et al., 2006; Hekkert and Leder, 2008) and have provided further insight into users' inferences between aesthetics and usability (e.g., Hartmann et al., 2008).

Hassenzahl (2004) wanted to further inquire into the nature of beauty in interactive products. He developed a theoretical model (Hassenzahl, 2005) that distinguishes between objective parameters, product quality perceptions and overall evaluations. Based on this model he understood beauty as "a high-level evaluative construct comparable to (but not identical with) other evaluative constructs, such as goodness or pleasantness" (Hassenzahl, 2004, , p. 323) and perceived usability as a bundle of lower level judgments reflecting product quality perceptions. He distinguished between two quality perceptions: *pragmatic* and *hedonic*. Pragmatic quality, he argued, refers to the product's ability to support the achievement of behavioral goals (i.e. usefulness and ease-of-use). On the contrary, hedonic quality refers to the users' self; it relates to *stimulation*, i.e. the product's ability to stimulate and enable personal growth, and *identification*, i.e. the product's ability to address the need of expressing one's self through objects one owns. He further distinguished between two overall evaluative judgments: *goodness* and *beauty*. Contrary to (Tractinsky et al., 2000), he found minimal correlation between pragmatic quality, i.e. usability, and beauty. Beauty was found to be a rather social aspect, largely affected by identification; pragmatic quality, on the contrary, related to the overall judgment of goodness. In a similar vein, Tractinsky and Zmiri (2006) distinguished between satisfying and pleasant experience. They found perceptions of usability to be better predictors of satisfying rather than pleasant experience while perceptions of products' aesthetics to be better predictors of pleasant rather than satisfying experience. Hassenzahl's (2004) model of user experience has also been further supported by subsequent research (e.g., Mahlke, 2006; Schrepp et al., 2006; Van Schaik and Ling, 2008).

Mahlke and Thüring (2007) provided a comprehensive framework linking product quality perceptions to emotional reactions and overall evaluative judgments. Their findings supported Hassenzahl's 2004 distinction between goodness and beauty, with goodness relating primarily to instrumental qualities, e.g. usefulness and ease-of-use, and beauty relating primarily to non-instrumental qualities such as the visual aesthetics and haptic quality (Mahlke, 2006). Desmet (2002), grounded on Russell's (1980) model of affect, developed a tool for measuring emotional responses to products and established a framework that relates aesthetic response to meaning (see Desmet and Hekkert, 2007). Fenko et al. (2009) studied how the dominance of different sensory modalities such as vision, audition, touch, smell and taste develops over different phases in the adoption of the product such as when choosing the product in the shop, during the first week, after the first month, and after the first year of usage.

1.2.2 Holistic Approaches

Holistic approaches are rooted in pragmatist philosophy and phenomenology. They criticize reductionist approaches in that they reduce the complexity and richness of user experience to "a set of manipulable and measurable variables" and impose "abstract models and classifications onto rich and complex models like affect and emotion" (Wright and Blythe, 2007) (see also Hassenzahl, 2008). Similarly, Suri (2002) argues that "measurement, by its nature, forces us to ignore all but a few selected variables. Hence, measurement is useful when we are confident about which variables are relevant". She argues that designers are concerned about developing new products and for new contexts and thus no such understanding exists about how design attributes and contextual details interact in given contexts, and proposes alternative methods, such as that of *experience narratives*, for inquiring into how product meaning and value emerges in given contexts.

From a theoretical point of view, holistic approaches have contributed a number of frameworks describing how experience is formed, adapted, and communicated in social contexts.

Forlizzi and Ford (2000) provided a framework that attempts to describe how experience transcends from unconsciousness to a cognitive state and finally becomes "an experience", something memorable that can also be communicated in social interactions. They identified four modes or dimensions of experiencing: sub-consciousness, cognition, narrative and storytelling. Sub-consciousness represents fluent experiences that do not compete for our attention. Cognition represents experiences that require our attention, e.g. learning to use an unfamiliar product. Narrative represents "experiences that have been formalized in the users' head: ones that force us to think about and formulate what we are doing and experiencing". Forlizzi and Ford (2000) suggest that a product's set of features and affordances offers a narrative of use. Storytelling, represents the subjective side of experience: "a person relays the salient parts of an experience to another, making the experience a personal story". Forlizzi and Ford (2000) argue that through these particular sense making users attach meaning and personal relevance to situation, "creating life stories and stories of product use". Forlizzi and Ford (2000) subsequently identify ways of shifting across these four modes of experiencing. One might migrate from a cognitive to a sub-conscious experience, for instance by learning how to use a product. Reversely, a fluent experience may shift to a cognitive if a user encounters something unexpected in her interaction with the product and is forced to think about it. A narrative experience can shift to a cognitive one when one "is forced to challenge his own thinking that has been solidified in her perceptions, attitudes, and beliefs". An experience might also shift from a sub-conscious state to story-telling as she "schematizes it, communicates it and add levels of meaning".

Forlizzi and Battarbee (2004) modified this framework to include the concept of co-experience proposed by Battarbee (2003). Battarbee and Koskinen (2005), further elaborated on the social mechanisms that lift or downgrade experiences as they participate in people's social interactions. They identified three distinct mechanisms: *lifting up experiences, reciprocating experiences, rejecting and ignoring experiences*. First, people may "lift things from the stream of events", considering them as meaningful enough to be communicated in social settings. Secondly, recipients of communicated experiences may acknowledge the described experience as personally relevant and respond to it by telling their own, similar experiences. Finally, experiences communicated in social settings may be rejected or downgraded by others, eventually altering the dominance of the given experience for the person who chose to communicate it.

McCarthy and Wright (2004) distinguished between four threads of experience: *compositional, sensual, emotional,* and *spatio-temporal*. The compositional thread concerns the way that different elements of experience form a coherent whole. It refers to "the narrative structure, action possibility, plausibility, consequences and explanations of actions". The sensual thread relates to "the concrete, palpable, and visceral character of experience that is grasped pre-reflectively in the immediate sense of a situation". The emotional thread refers to value judgments (e.g., frustration and satisfaction) that ascribe importance to other people and things with respect to our needs and desires". Lastly, the spatio-temporal thread "draws attention to the quality and sense of space-time that pervades experience". McCarthy and Wright (2004) pinpoint that while these are positioned as distinct components of experience they should be seen as intrinsically connected with each other.

Next to the four threads of experience, McCarthy and Wright (2004) described how sense-making takes place in the development of experience by decomposing it into six processes: *anticipating, connecting, interpreting, reflecting, appropriating,* and *recounting*. Anticipation refers to users' expectations and imagined possibilities that are grounded in prior experience. In connecting, users make an instant judgments referring to the immediate, pre-conceptual and pre-linguistic sense of a situation. In interpreting, users work out what's going on and how they feel about it. In reflecting users examine and evaluate what is happening in an interaction and the feelings of frustration or pleasure that are part of the experience. In appropriating, users evaluate how the new experience relates to prior experiences, and in recounting, users communicate the experienced situation to others and reinterpret the experience as it participates in storytelling.

1.3 Diversity in User Experience

While one may quickly note that holistic approaches emphasize the uniqueness of experience across different situations and people, both approaches to user experience have readily accepted that diversity in users' experiences is prevalent. A wealth of empirical studies, grounded on reductionism, acknowledge and tackle diversity empirically (e.g. Hassenzahl and Ullrich, 2007; Mahlke and Lindgaard, 2007).

This section introduces the notion of diversity in user experience. We introduce a framework of diversity in subjective judgments and identify four different sources of diversity in users' experiences with interactive products. Only later, in section 1.4 we will introduce the methodological differences between the two approaches, reductionist and holistic, in accounting for diversity.

1.3.1 A Framework of Diversity in Subjective Judgments

Diversity was readily accepted in the HCI field as a key issue. Not all users like the same things and different product qualities suffice in different situations (e.g. Cooper, 1999). But, other fields have been constructed on the assumption of homogeneity across different individuals. In the field of psychophysics, for example, the principle of *homogeneity of perception* states that different participants will more or less agree on perceptual judgments such as how much noise, or blur, an image contains, or how much friction, or inertia, one may find in a haptic control. This assumption has been instrumental in the development of respective statistical techniques; for instance, Multi-Dimensional Scaling (Green et al., 1989; Martens, 2003), motivated by this principle, assumes that judgments of different individuals may be visualized in a shared K-dimensional configuration of stimuli, for which the coordinates of the stimuli in the configuration space along different axes can be monotonically related to the observed attribute ratings of the participants.

It seems natural to accept that while different individuals might agree on low-level perceptual judgments, e.g. friction in a haptic control, a relative disagreement would be found as one moves to more cognitive judgments, e.g. the degree to which this haptic control is perceived as playful. Indeed, in an exploratory study we tried to inquire into whether people agree on *product character* judgments (Janlert and Stolterman, 1997) of a physical rotary knob. A project done by Bart Friederichs in his Master's studies, in collaboration with BMW, aimed to prove that physical controls, i.e. rotary knobs, can be designed so that they comply with the personality of a given car. But, how much do different users agree on high level product character judgments of haptic controls? Eighteen participants experienced fifteen different haptic scenes programmed in a haptic feedback knob and rated them in terms of four different bipolar scales: *Dynamic - Static, Agile - Plump, Playful - Serious,* and *Luxurious - Austere.* One may quickly note that the last two judgments, those of playfulness and luxury, are higher level judgments, allowing for more disagreement on what one considers playful or luxurious. We applied a uni-dimensionality test, by applying two Multi-Dimensional Scaling solutions on the four ratings: a 1D and a 2D. A χ^2 test, showed that the first two judgments, which we assumed as lower level, perceptual judgments, were indeed more uni-dimensional, i.e. they displayed higher consistency across different individuals.

In the field of design, Csikszentmihalyi and Rochberg-Halton (1981) asked individuals to select personally relevant objects located in their homes and describe what makes them special. Csikszentmihalyi and Rochberg-Halton (1981) found that the value of such objects did not lie in some objectively defined quality, e.g. uniformly appreciated aesthetics, but to the personal meaning that people attached to these objects and how these participated in their social lives and creation of self-identity. These results suggest that while we may all agree in perceptual judgments, e.g. color of a given product, these judgment have far less power in predicting our preferences in comparison to higher level judgments, e.g. its beauty. Hofstede (2001) suggested that human perceptions may exist at three different levels: a) some might be uniquely personal, having significance for an individual because of his or her own associations; b) others might have significance to a specific social or cultural group through shared meaning; and c) others are universal, related to human nature at an innate or fundamental level.

In Karapanos et al. (2008b) we argued that diversity may exist at two different stages in the formation of an overall evaluative judgment (see figure 1.1). Perceptual diversity lies in the process of forming product quality perceptions (e.g. novel, easy to use) on the basis of product features. For instance, different individuals may infer different levels on a given quality of the same product, e.g. disagree on its novelty. Evaluative diversity lies in the process of forming overall evaluations of the product (e.g. good-bad) on the basis of product quality perceptions. For instance, different individuals may form different evaluative judgments even while having no disagreement on the perceived quality of the product, e.g. both might think of it as a novel and hard-to-use product, but they disagree on the relative importance of each quality.

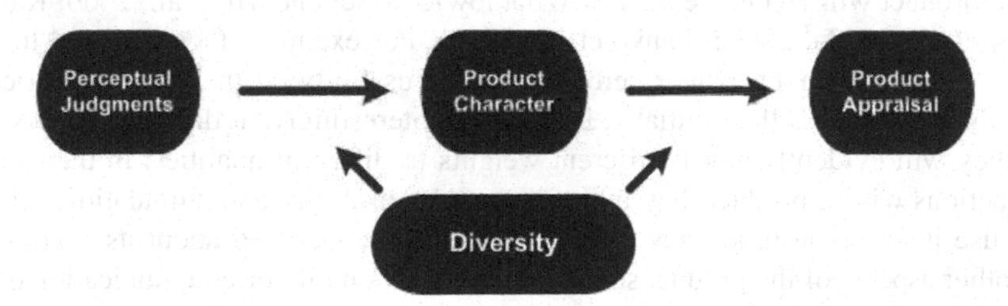

Fig. 1.1 A modification of Hassenzahl's 2005 framework, highlighting diversity at two stages in forming overall judgments about products.

Considering the second stage, i.e. evaluative diversity, one might assume a certain universal hierarchical structure on the relative importance of different qualities. For instance, Jordan (2000) drew on Maslow's (1946) theory of human needs to propose a fixed hierarchical structure of the relative importance of functionality, usability and pleasure in the context of Human-Computer Interaction. According to this, a product has to provide useful and usable functionality before hedonic aspects, such as beauty and stimulation, can take effect. Similarly to the field of psychology where Maslow's theory was readily adopted while lacking empirical evidence, Jordan's fixed hierarchy has become widely popular in the field of user experience, but, to our knowledge, no empirical studies have attempted to confirm or disprove the framework. Contrary to Jordan (2000), Hassenzahl (2006) assumes the importance of these different qualities to vary with several contextual factors.

1.3.2 Four Sources of Diversity in User Experience

In the field of user experience, a wealth of empirical studies have tried to tackle how the relative importance of different product qualities on users' overall evaluations and preferences, is modulated by a number of contextual aspects. In figure 1.2 we highlight four different sources of diversity that user experience research has been concerned with.

First, individual differences (e.g. human values Schwartz, 1992) moderates the importance individuals attach to different qualities of an interactive product (e.g. Desmet et al., 2004; Karapanos and Martens, 2007); while some might prefer playful and stimulating products, others might value simplicity and austerity. Second, the type of the product matters (e.g. Jordan and Persson, 2007); while a playful interaction might be crucial for the success of a computer game, the same quality might be perceived as inadequate for professional software. Third, even for the same product, the way individuals use it differs across situations and this impacts the importance that they attach to different qualities (e.g. Mahlke and Lindgaard, 2007; Hassenzahl and Ullrich, 2007; Hassenzahl et al., 2008); the same mobile phone could be used for exploring the available ring tones or to make an emergency call.

A fourth aspect, which is mostly overlooked so far, is the systematic change of experience over time. As individuals use a product, their perception of the qualities of the product will change (e.g. von Wilamowitz Moellendorff et al., 2006; Karapanos et al., 2008a, 2009c; Fenko et al., 2009). For example, they get used to it, which eventually changes their perception of its usability; at the same time it excites them much less than initially. Even more interestingly, at different phases of use they will evidently attach different weights to different qualities. In their first interactions with a product they may focus on its usability and stimulation. After they use it for some time, they might become less concerned about its usability, and other aspects of the product such as novel functionality or communication of a desired identity towards others become more important.

All these factors, the individual, the product, the situation, and time modify the importance of the qualities for a satisfying experience with an interactive product. In this view, Jordan's (2000) hierarchy of consumer needs could be seen as a particular, context-dependent prioritization of needs (Hassenzahl, 2006).

This section introduced the notion of diversity, which mostly concerns in the user experience field the relative dominance of different product qualities while forming overall judgments about interactive products. The next section introduces the methodological debate in accounting for diversity between the two dominant

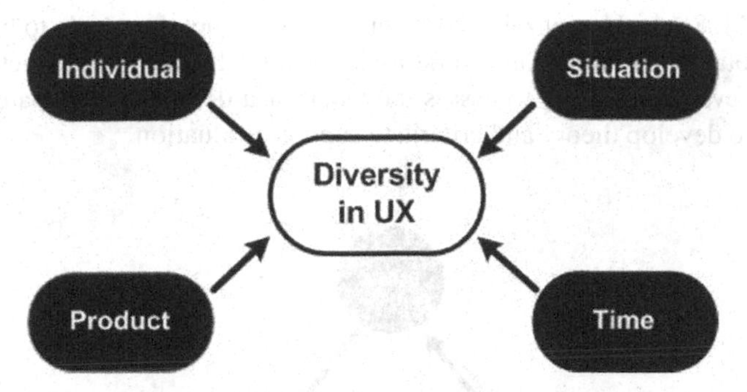

Fig. 1.2 Four different sources of diversity in user experience. These are assumed to modulate the dominance of different product qualities on users' experience and overall evaluative judgments.

approaches in user experience research, it discusses the limitations of each approach, and argues for a hybrid methodological paradigm that shares values from both approaches, the holistic and the reductionist.

1.4 Methodological Issues in Accounting for Diversity

Reductionist approaches typically employ psychometric scales (e.g. Likert, 1932; Osgood et al., 1957) in measuring latent psychological constructs that the researchers consider as relevant for a certain context. Such approaches are grounded on the assumption that people may summarize experiences in overall evaluative judgments (Hassenzahl, 2008). When a product is associated with a certain experience, the value of the experience will be partially attributed to the product.

Proponents of the holistic paradigm criticize reductionist approaches for not capturing "grey areas and moment-by-moment decision making" (Wright and Blythe, 2007). Instead, they propose holistic and situated techniques that are motivated by the need to establish an empathic relation with the user (see Wright and McCarthy, 2008). In one of the most popular techniques, the *cultural probes* (Gaver et al., 1999), participants are typically provided with a camera and a set of abstract objects to inspire them in capturing their experiences, dreams and aspirations related to a given topic of interest. Suri (2002) proposed the elicitation of short essays, i.e. *experience narratives*, in understanding the frustration and enjoyment that is experienced while interacting with products. Blythe et al. (2002) proposed a technique called *technology biographies* in which participants are asked to reminisce how their relationship with technology has changed through their lives. In the creation of*Anticipation and reflection interviews*, (Blythe et al., 2006) acknowledge that experience is not limited to a given interaction instance but extend to the process of forming expectations of and reflecting on experiential episodes.

One might argue that these two approaches have distinct goals, being *evaluative* or *inspirational*. In this sense, holistic approaches to evaluation serve to inspire design solutions (Hassenzahl, 2008); evaluation is here understood as idea generation

(Hornbæk, 2008). As Hassenzahl (2008) puts it "[designers] are able to build ideas from anecdotal observations and loose associations". In contrast, reductionist approaches to evaluation serve to assess the value of a design, to compare multiple designs, or to develop theory and criteria to support evaluation.

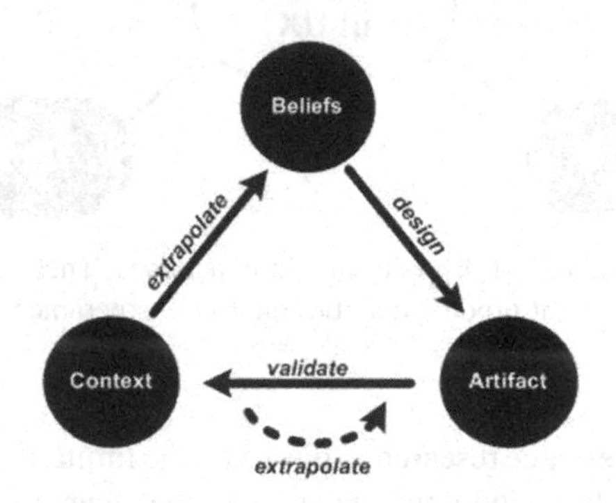

Fig. 1.3 A simplified framework of the design process, derived from (Martens, 2009b, personal communication). According to this framework, the designer forms beliefs about the potential consequences of use of a given product in a given context, grounded on empirical knowledge. These beliefs are in turn externalized to artifacts and the artifacts are validated on a number of criteria that have been derived from prior beliefs, or empirical insights from the previous design iteration.

We argue against this distinction between evaluative and inspirational goals in product evaluation. It is argued that current evaluation practices that embrace this distinction lead to inefficacy in the evaluation of interactive products. This is reflected in figure 1.3 which depicts a simplification of a framework of the design process initially proposed by (Martens, 2009b, personal communication). The framework identifies three distinct activities in an iterative design process: a) extrapolating knowledge from context and forming beliefs about the potential consequences of use of a given product within a given context, b) transforming (implicit) beliefs to explicit knowledge through the design of artifacts, and c) validating the designed artifacts in context.

According to this view, while holistic approaches serve to extrapolate knowledge in forming or adapting the designer's beliefs, reductionist approaches serve to validate the designed artifacts using a number of criteria that have been derived from prior beliefs, or empirical insights from the previous design iteration. This leads to a number of limitations in the evaluation process. First, reductionist approaches to evaluation through pre-defined measurement scales may miss potentially relevant concepts due to a failure of researchers to recognize their relevance, but may also result in less meaningful information when participants cannot interpret the questions with respect to their own context. Second, holistic approaches such as the experience narratives, provide rich information in all aspects that surround an experience but may create a risk of focusing on idiosyncratic experiences while failing to estimate its dominance and probability of occurrence over the total population of users.

In combination, rich situated insights derived from holistic methods may inform the validation of artifacts while reductionist approaches to evaluation may quantify the importance of a given experience and thus minimize the risk of overemphasize interesting but rare experiences.

We argue for a missing link between *validation* and *extrapolation* (see figure 1.3). Below, we describe how this is addressed in relation to the two research foci of this manuscript: understanding interpersonal diversity in users' responses to conceptual designs, and understanding the dynamics of experience over time.

1.4.1 Understanding Interpersonal Diversity through Personal Attribute Judgments

Den Ouden (2006) revealed that the majority of soft reliability problems related to the concept design phase and were particularly rooted in design decisions relating to the product definition. This insight suggests that design decisions made early in the design process may not be adequately grounded on empirical user insights.

Traditional approaches to measuring users' responses to artifacts lie in the a-priori definition of the measures by the researchers. This approach is limited in at least two ways when one is concerned with capturing the richness of and diversity in user experience. First, the a-priori definition of relevant dimensions is inherently limited as researchers might fail to consider a given dimension as relevant, or they might simply lack validated measurement scales, especially in developing fields such as that of user experience where radically new constructs are still being introduced. Secondly, one could even wonder whether rating a product on quality dimensions that are imposed by the researcher is always a meaningful activity for the user, for example when the user does not consider a quality dimension as relevant for the specific product. There is increasing evidence that users are often unable to attach personal relevance to the statement provided in psychometric scales due to a failure to recall experiential information that relates to the statement or due to lengthy and repetitive questioning. Larsen et al. (2008b) reviewed a number of studies employing psychometric scales in the field of Information Systems. They found for the majority of studies the semantic similarity between items to be a significant predictor of participants' ratings $(.00 < R^2 < .63)$. In such cases, they argued participants are more likely to have employed *shallow processing* (Sanford et al., 2006), i.e. responding to surface features of the language rather than attaching personal relevance to the question.

An alternative approach to predefined questionnaires lies in a combination of structured interviewing, that aims at eliciting attributes that are personally meaningful for each partcipant, with a subsequent rating process performed on the attributes that were elicited during the interview. This approach aims at increasing the diversity and relevance to the individual of concepts that are measured, thus resulting in richer insights. However, the techniques required for the quantitative analysis of such information become significantly more complex. We propose two techniques for the analysis of personal attribute judgments.

1.4.2 Understanding the Dynamics of Experience through Experience Narratives

Product evaluation practices have traditionally been focusing on early interactions. As a result, we have been mostly concerned about product qualities relating to the initial use. Den Ouden et al. (2006), however, highlighted that the reasons for product returns span a wider range of aspects than just problems related to their learnability and usability. Moreover, a number of recent trends are highlighting the importance of longitudinal evaluation practices. First, legislation and competition within the consumer electronics industry has resulted in an increase in the time-span of product warranties. This has resulted in an alarmingly increasing number of products being returned on the basis of failing to satisfy users' true needs (Den Ouden et al., 2006). Secondly, products are increasingly becoming service-centered. Often, products are being sold for lower prices and revenues are mainly coming from the supported service (Karapanos et al., 2009c). Thus, the overall acceptance of a product shifts from the initial purchase to establishing prolonged use.

Traditional approaches in the study of dynamics of experience over time typically employ validated measurement and structural models across different phases in the adoption of a system (e.g. Venkatesh and Davis, 2000; Venkatesh and Johnson, 2002; Kim and Malhotra, 2005). For instance, Venkatesh and Davis (2000) employed the Technology Acceptance Model (Davis et al., 1989) at three instances in the adoption of an information system at a workplace: before the introduction of the system (inquiring into users' expectations), right after the introduction of the system, and three months after the introduction. These approaches, while being widely validated and well-cited in the field of Information Systems, are hindered by a number of limitations, at least in developing fields such as that of user experience.

An assumption inherent in this approach is that the relevant latent constructs remain constant, but their perceived value and relative dominance might change over time. But, especially in developing fields such as user experience, substantial variations might occur over time even in what constructs are relevant to measure. Some constructs, e.g. novelty, might cease to be relevant while others, such as supporting daily rituals, enabling personalization, and communicating a positive image about one's self (see Karapanos et al., 2009c), that were not evident in studies of initial use might become critical for the long-term acceptance of a product. Firstly, this might challenge the content validity of the measurement model as relevant latent constructs might be omitted. This is often observed in studies of user acceptance and user experience where the a-priori defined constructs account for a limited amount of the variance in the predicted variable, being it preference judgments, dissimilarity judgments or attitude towards behavior (see Venkatesh et al., 2003). Secondly, it may also lead to distorted data as individuals might fail to interpret the personal relevance of a given scale item to their own context, for instance when a construct ceases to be relevant over prolonged use. Last, such approaches provide rather limited insight into the exact reasons for changes in users' experiences. They may, for instance, reveal a shift in the dominance of perceived ease-of-use and perceived usefulness on intention to use a product (e.g. Venkatesh and Davis, 2000), but provide limited insight to the exact experiences that contributed to such changes.

An alternative approach for the measurement of the dynamics of experience over time relies on the elicitation of idiosyncratic self-reports of one's experiences with a product, i.e. *experience narratives*. Each narrative provides rich insights into a given experience and the context in which it takes place. Moreover, generalized knowledge may also be gained from these experience narratives. Such generalized knowledge may be reflected in questions like: how frequent is a certain kind of experience, what is the ratio of positive versus negative experiences and how does this compare to competitive products, how does the dominance of different product qualities fluctuate over time and what should be improved to motivate prolonged use?

This manuscript makes two methodological contributions in this research problem. First, it highlights the labor-intensive nature of longitudinal studies, and proposes an alternative approach that relies on the elicitation of one's experiences with a product from memory. iScale, a tool designed with the aim of increasing users' effectiveness and reliability in recalling their experiences is theoretically grounded and empirically validated. Second, it proposes a computational approach that aims at supporting the researcher in the qualitative analysis of experience narratives. The proposed approach addresses two limitations of traditional qualitative analysis practices. First, qualitative analysis is a labor intensive activity which becomes increasingly a concern when qualitative data may be elicited from a large number of participants as in the case of iScale. Second, qualitative analysis has been shown to be prone to researcher bias as humans often rely on heuristics in forming judgments about the relevance or similarity of two or more data instances (Kahneman et al., 1982). The proposed approach aims at supporting the researcher through semi-automating the process of qualitative coding, but also minimizes the risks of overemphasizing interesting, but rare experiences that do not represent users' typical reactions to a product.

1.5 Manuscript Outline

We argue for a hybrid paradigm between reductionist and holistic approaches to evaluation. We propose two techniques, one grounded in *personal attribute judgments* and one in *experience narratives*. Both developed techniques aim at increasing the richness and diversity in obtained information while trying to create different levels of granularity of insight, thus enabling the researcher to move between abstracted, generalized insight and concrete, idiosyncratic and insightful information.

Part I
Understanding Interpersonal Diversity through Personal Attribute Judgments

Chapter 2 highlights the limitations of standardized psychometric scales and introduces personal attributes judgments. It introduces attribute elicitation techniques and in particular, the Repertory Grid Technique (RGT). It argues that the true value of RGT is in *quantifying rich qualitative insights* and highlights the limitations of relevant statistical techniques that are typically employed in the analysis of Repertory Grid data. It presents an initial Multi-Dimensional Scaling procedure that aims at identifying diverse views in Repertory Grid data. The procedure identifies distinct user groups in a sample population and derives a two-dimensional view for each respective user group. The technique is presented through a case study where users' views on a set of product concepts were contrasted to the ones of designers.

Chapter 3 presents a second Multi-Dimensional Scaling procedure that aims at identifying diverse views even within single individuals. The technique is applied on an existing dataset (Heidecker and Hassenzahl, 2007). It is illustrated that the - traditional - averaging analysis provides insight to only 1/6th of the total number of attributes in the example dataset. The proposed approach accounts for more than double the information obtained from the average model, and provides richer and semantically diverse views on the set of stimuli.

Part II
Understanding the Dynamics of Experience through Experience Narratives

Chapter 4 presents two studies that inquired into how users experiences with interactive products develop over time. In the first pilot study, grounded on reductionism, we asked participants to rate a novel product during the first week as well as after four weeks of use. In the second study six participants were followed after the purchase of a novel product and elicited rich experience narratives over a period of one month.

Chapter 5 presents iScale, a survey tool that aims at eliciting users' experiences with a product in the form of experience narratives. iScale employs sketching in imposing a process in the reconstruction of one's experiences from memory. The chapter motivates the development of two distinct versions of iScale which were grounded in two opposing theoretical approaches to reconstructing one's emotional experiences from memory. Finally, it presents two studies that compared the two different versions of iScale with traditional methods.

Chapter 6 proposes a semi-automated technique for the content analysis of experience narratives. The technique combines traditional qualitative coding procedures (Strauss and Corbin, 1998) with computational approaches for assessing the semantic similarity between documents (Salton et al., 1975). This results in an iterative process of qualitative coding and visualization of insights which enables to move quickly between high-level generalized knowledged and concrete and idiosyncratic insights.

Chapter 7 concludes the research described in this manuscript by reflecting on its contributions and sketching directions for future research.

Chapter 2
Personal Attribute Judgments

Abstract. Traditional approaches to measuring users' responses to artifacts lie in the a-priori definition of the measures by the researchers. This chapter highlights the limitations of such approaches that employ standardized psychometric scales and introduces personal attributes judgments. It introduces attribute elicitation techniques and in particular, the Repertory Grid Technique (RGT). It argues that the true value of RGT is in *quantifying rich qualitative insights* and highlights the limitations of relevant statistical techniques that are typically employed in the analysis of repertory grid data. An initial Multi-Dimensional Scaling (MDS) procedure that aims at identifying diverse views in Repertory Grid data is proposed. The procedure identifies distinct user groups in a sample population and derives a two-dimensional view for each respective user group. The technique is presented through a case study where users' views on a set of product concepts were contrasted to the ones of designers. The technique revealed differences not only between users and designers but also between designers of different professional background and role in the design team.

2.1 Introduction

Reductionist approaches to user experience evaluation are grounded on the assumption that people may summarize experiences in overall evaluative judgments (Hassenzahl, 2008). When a product is associated with a certain experience, the value of the experience will be partially attributed to the product. Such evaluations may be *substantive* (Hassenzahl, 2004), e.g. perceptions of a given product quality such as usability, or *verdictive*, e.g. overall evaluations of goodness, appeal, or beauty.

These are all *latent* constructs, in that they cannot be directly measured but instead, they are estimated through a number of different measures. For instance, the Unified Theory of Acceptance and Use of Technology (Venkatesh et al., 2003) measures a latent construct termed *performance expectancy* through a set of four items such as "I would find the system useful in my job" and "Using the system increases my productivity". These items are assumed to be measuring different facets of the

E. Karapanos: Modeling Users' Experiences with Interact. Syst., SCI 436, pp. 17–39.
springerlink.com

same latent construct (e.g. performance expectancy), at the same time being more or less uni-dimensional.

The development of psychometric scales is often described as a three-step process: *item generation, scale development,* and *scale evaluation* (Hinkin, 1995). The first step aims at enhancing the content validity of the questionnaire (i.e. that a complete coverage of the domain of interest is obtained through the proposed items); the latter two steps aim at enhancing the convergent and discriminant validity of the questionnaire (i.e. that each item correlates highly with other items that attempt to measure the same latent construct, and weakly with items that attempt to measure different latent constructs).

Once a substantial set of latent constructs have been developed for a given field, questionnaires may be used by researchers and practitioners to assess the value of products. Using validated questionnaires, one can measure how two or more products compare on a given quality dimension (e.g. trust), or compare two different generations of the same product to assess the impact of the redesign process.

Proponents of the holistic approach in user experience criticize the use of psychometric scales for their inability to capture the richness and diversity of experience (see Blythe et al., 2007). Below, we will try to expand on this criticism by highlighting two limitations of a-priori defined psychometric scales. We will then introduce the use of personal attribute judgments as a means to account for diversity in users' experiences with interactive products.

Firstly, measures that are a-priori defined are inherently limited in accounting for the users' perspective. Especially in developing fields such as that of user experience, where the development of new constructs is still in its infancy, researchers might fail in capturing a relevant experience dimension due to a failure in recognizing its importance or simply due to the absence of relevant measurement scales. This issue has been repeatedly highlighted in studies of user acceptance of information systems that employ pre-defined measurement and structural models such as the Technology Acceptance Model (TAM) (Davis et al., 1989). A number of studies have reported limited predictive power of the Technology Acceptance Model, in some cases accounting for only 25% of the variance in the dependent variable (Gefen and Straub, 2000). Lee et al. (2003) reported that *"the majority of studies with lower variance explanations did not consider external variables other than original TAM variables"*. A typical case is illustrated in figure 2.1 which displays a two-dimensional visualization of the perceived dissimilarity of three systems (van de Garde-Perik, 2008). In the mentioned study, users were asked to judge the overall dissimilarity of the three systems as well as rate the systems on a number of pre-defined dimensions such as the perceived trust, risk, usefulness and ease of use. The configuration of the three stimuli is derived by means of Multi-Dimensional Scaling on the original ratings of dissimilarity, while the latent constructs are fitted as vectors in the two-dimensional space by means of regression. While a number of insights may be derived from this visualization, one may note that systems 3 and 1 are clearly differentiated in terms of their overall dissimilarity while none of the predefined attributes can explain this dissimilarity. In other words, the measurement

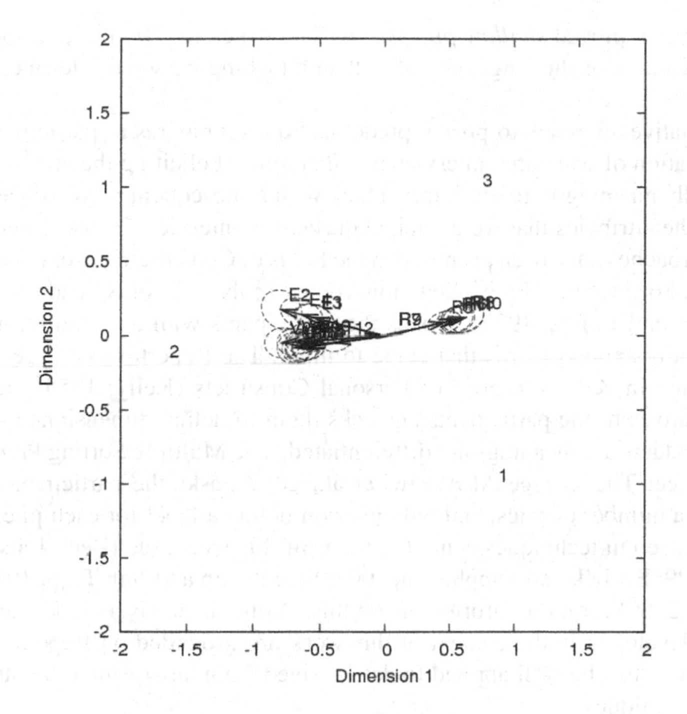

Fig. 2.1 Multi-Dimensional Scaling on dissimilarity ratings for three products with latent constructs fitted as vectors. Note that the latent constructs employed in this study fail to explain the dissimilarity between products 1 and 3.

model fails to capture the full set of qualities that are meaningful to the participants of the study.

Secondly, one could even wonder whether rating a product on quality dimensions that are imposed by the researcher is always a meaningful activity for the user, for example when the user does not consider a quality dimension as relevant for the specific product. A basic assumption that underlies the rating process is that the user is able to understand, interpret and position a given statement within her own context. For instance, if the user is asked to rate the product in terms of its usefulness, she is expected to interpret usefulness in her own context, to identify what aspects of her experience with the given product contribute to perceptions of usefulness and summarize them in an overall index of usefulness. First, this process is an inherently idiosyncratic process; thus, different users might reflect on different facets of the construct. Secondly, users might fail in certain occasions to interpret the personal relevance of the statement in their own context, e.g. due to a failure to recall experiential information that relates to the statement or due to lengthy and repetitive questioning. Larsen et al. (2008b) reviewed a number of studies employing psychometric scales in the field of Information Systems. They found for the majority of studies the semantic similarity between items to be a significant predictor of participants' ratings *($.00 < R^2 < .63$)*. In such cases, they argued participants are more likely to have employed *shallow processing* (Sanford et al., 2006), i.e. responding to surface features of the language rather than attaching personal relevance to the question.

An alternative approach to posing predefined questionnaires to participants lies in a combination of structured interviewing, that aims at eliciting the attributes that are personally meaningful for each individual, with a subsequent rating process performed on the attributes that were elicited during the interview. Many different interview approaches have been proposed in the fields of Constructivist and Economic Psychology. For instance, Free Elicitation, rooted in theories of Spreading Activation (Collins and Loftus, 1975), probes the participants with a stimulus and asks them to rapidly express words that come to mind. The Repertory Grid Technique (RGT), rooted in Kelly's Theory of Personal Constructs (Kelly, 1955), provides three alternatives to the participants and asks them to define dimensions in which the three products are meaningfully differentiated. The Multiple Sorting Procedure, rooted in Facet Theory (see Al-Azzawi et al., 2007), asks the participant to sort products in a number of piles, and only later on define a label for each pile. Comparing the different techniques is not the focus of this work; see (Bech-Larsen and Nielsen, 1999; Breivik and Supphellen, 2003; Steenkamp and Van Trijp, 1997; van Kleef et al., 2005) for more information on this. While the analysis procedures that have been developed in the context of this work are grounded on Repertory Grid data, they may also be well applied to data derived from any of the other attribute elicitation techniques.

2.2 The Repertory Grid Technique

The RGT is one of the oldest and most popular attribute elicitation techniques. It originates from Kelly's Personal Construct Theory (PCT) (Kelly, 1955, 1969) which suggests that people form idiosyncratic interpretations of reality based on a number of dichotomous variables, referred to as personal constructs or attributes. A personal construct is a bi-polar similarity-difference judgment. For example, when we meet a new person we might form a construct *friendly-distant* to interpret her character. In this process we perform two judgments: one of similarity and one of dissimilarity. Both judgments are done in comparison to reference points: people that we regard as friendly or distant.

To elicit the idiosyncratic attributes of each individual, the RGT employs a technique called *triading*, where the participant is presented with three products and is asked to *"think of a property or quality that makes two of the products alike and discriminates them from the third"* (Fransella et al., 2003). Once a bipolar construct is elicited, the researcher may further probe the participant to elaborate on the construct through the *laddering* and the *pyramiding* techniques (see Fransella et al., 2003). Laddering seeks to understand what motivates a given statement and thus ladders up in an assumed means-ends-chain (Gutman, 1982) towards more abstract qualities of the stimuli; in laddering the researcher typically asks the participant whether the mentioned quality is positive or negative, and subsequently probes the participant to motivate its importance, e.g. "why is expressiveness important to you?" Pyramiding, on the other hand, also known as negative laddering, seeks to understand the lower level attributes that make up for a given quality; in pyramiding the researcher asks the participant to elaborate on what makes the given product to be characterized with the respective attribute, e.g. "what makes this product more [easy to use]?". This process can be repeated for all possible combinations of products and until no new attributes arise. The result is a list of attributes that the specific individual uses to differentiate between a set of products. The attributes may then be employed in rating scales, typically Semantic Differentials (Osgood et al., 1957), and each participant rates the set of products on her own elicited attributes. Participants' ratings are subsequently analyzed with exploratory techniques such as Principal Components Analysis (PCA) or Multi-Dimensional Scaling (MDS).

With the recently increased interest in user experience (e.g. Hassenzahl and Tractinsky, 2006), the RGT has become popular in the field of HCI. Hassenzahl and Wessler (2000) employed the RGT to evaluate the outcome of parallel design and analyze the perceived character of websites (Hassenzahl and Trautmann, 2001). Fallman and Waterworth (2005) elicited users' experiences with mobile technology devices, while Davis and Carini (2004) explored player's experience of fun in video games. Hertzum et al. (2007) studied the differences between designers' and users' perceptions for three diverse cultural settings while others have used the RGT as an ethnographic method in understanding how individuals organize their documents and communications (e.g. Szostek et al., 2008; Bondarenko and Janssen, 2009). It, thus, becomes evident that an increasing number of researchers in HCI, emphasize the idiosyncratic nature of subjective judgments on the quality of interactive products.

2.3 The Quantitative Side of Repertory Grid - Some Concerns

While the RGT has become popular in the field of Human-Computer Interaction and User Experience, one may note several problems in the application of RGT in practice, at least from a quantitative perspective. Below we highlight these issues and introduce the contribution of our work in the analysis of Repertory Grid data.

2.3.1 Are We Really Interested in Idiosyncratic Views?

The RGT originates from clinical psychology where the emphasis is on an individual's perceptual and cognitive processes. In the field of user experience, however, the interest is not in the idiosyncratic views of an individual but rather on some more-or-less homogeneous groups of individuals. Due to this focus on idiosyncrasy within the field of personal construct psychology, interpersonal analysis of repertory grid data has received very little attention. We have come to realize two inadequacies of the RGT when used for interpersonal analysis.

First, the interpersonal classification of attributes is often performed purely on semantic grounds, i.e., without testing congruence on the rating scores. Such a practice does not take properly into account the diverse ways in which individuals construe attributes to refer to internal concepts (Kelly, 1955). It is only when two attributes agree both on semantics and in terms of ratings that one might be confident that the two participants refer to the same concept. One could argue that in certain cases two attributes refer to the same concept, but different participants value products differently on the same concept. For instance, one participant might consider product A more easy to use than product B with a second participant considering the inverse, sharing however the same notion of ease of use. Even in this extreme case, one should not group two attributes with diverse ratings. As it will become apparent in the next chapters, we place ratings at higher importance than semantics, i.e. two attributes are checked for semantics only when there is evidence that, within the limited set of stimuli, they (quantitatively) represent the same internal concepts.

Second, techniques such as Principal Components Analysis (PCA) or Multi-Dimensional Scaling (MDS) that are typically employed in the analysis of repertory grid data assume homogeneity in the way people perceive the stimuli offered to them. This is sometimes referred to as the *principle of homogeneity of perception* (Martens, 2003). To our knowledge, however, all RGT approaches up to date have been employing averaging techniques for the quantitative analysis of personal attribute judgments. We believe this to be due to a lack of more advanced techniques that can account for diversity in users' subjective judgments, eventually undermining the core value of the RGT, i.e. to account for diversity in individuals' subjective judgments. We propose a quantitative, exploratory MDS procedure that aims at accounting for the diverse views that one or more individuals may have on a set of products. It will be demonstrated that even single participants can handle more than one view on a set of stimul i. It will be shown that by averaging interesting views are overlooked due to majorization bias.

2.3.2 On Bipolarity

A central notion in the RGT is the bipolarity of the idiosyncratic constructs (i.e. attributes). Kelly, in his theory of Personal Constructs, postulated that individuals perceive the world around them through the construction of dichotomous constructs. It is our experience, however, that participants often need to be probed in order to derive a truly bipolar attribute. This raises concerns with respect to whether individuals actually do think in bipolar terms. Lyons (1977) posited that *"categorizing experience in dichotomous contrasts is a universal human tendency which is only secondarily reflected in language"*. He identified three different types of bipolarity: *negation* (i.e. practical-impractical), *opposition* (i.e. professional - amateurish) and *non-contiguous*, where the opposite pole does not constitute a negation or linguistic opposition (i.e. easy - powerful) (c.f. Yorke, 2001).

In a meta-analysis of a RGT study on early concept evaluation (Karapanos and Martens, 2007), we found that the majority of the elicited constructs (67%) were negation constructs, while 17% were opposition constructs and only 16% were non-contiguous. This deviates substantially from what was observed in a study by

Table 2.1 Percentages of attributes types from Karapanos and Martens (2007) (Study 1) and Hassenzahl and Wessler (2000) (Study 2) studies.

Type	Study 1	Study 2
Negation	67%	35%
Opposition	17%	26%
Non-contiguous	16%	39%

Hassenzahl and Wessler (2000) where non-contiguous constructs constituted the major category (39%) while negation accounted for 35% of the constructs and opposition for the remaining 26%. This observed discrepancy is likely to have been influenced by two aspects: a) the instructions of the experimenter (our study had a more evaluative character aiming at eliciting attributes that can be validly used as semantic differential scales, while the latter study's primary goal was to inform design through rich qualitative accounts) and b) the fidelity of the prototypes (in our study early concepts were communicated in sketched scenarios, while in the latter study users interacted with working prototypes).

This highlights a problem rooted in the dual nature of the RGT: it aims at eliciting rich qualitative accounts which can also be quantified. Non-contiguous attributes provide insight into the relationships that individuals perceive between design qualities (i.e. beautiful - hard-to-use) and concrete product attributes (i.e. easy to use - has many buttons). They provide rich information to design. They are however inherently problematic when used in psychometric scales as the two poles do not underly a single uni-dimensional construct, and thus they will evidently elicit distorted ratings. In our experience we have seen cases where, during rating, participants cannot recall the context in which the attributes were elicited. When both poles of a bipolar construct are not (equally) evident to the participant, ratings may very well be based mostly on one of the poles. We would thus suggest that attributes should be validated by the participant before moving to the rating phase. In this attribute validation phase, participants can be asked to remove duplicate attributes and rephrase attributes when needed.

Negation and opposition bipolarity constitute the common practice in validated questionnaires. In a small study we attempted to explore the difference, if any, between negation and opposition bipolarity in rating scales. Our interest was to test the opposition hypothesis, i.e. ratings for the negative and positive pole should have a linear correlation of -1. Fourteen participants, all students at the department of Industrial Design, rated three concepts on two negation (i.e. secure-insecure, practical-impractical) and two opposition (i.e. standard-creative, amateurish-professional) attributes, using paired comparison scales. The two poles of each scale were split in two distinct scales; ratings on these two scales (e.g. secure and insecure) should approximate a correlation of -1. All attributes except the one referring to security were selected from Attracdiff2 (see Hassenzahl, 2004), a validated user experience questionnaire. Attributes had been translated to Dutch (Frens, 2006). These three attributes were also identified in a repertory grid study with the three concepts

Table 2.2 Amount of variance accounted for by the latent one-dimensional construct. Original attributes were in Dutch.

English (translated)	Dutch (original)	R^2
Negation		
Secure-Insecure	Veilig-Onveilig	0.88
Practical-Impractical	Praktisch-Onpraktisch	0.86
Opposition		
Creative-Standard	Creatief-Fantasieloos	0.75
Professional-Amateurish	Vakkundig-Als een leek	0.90

(described in the remaining of the paper), ensuring the content validity of the scales for the current study.

The opposition hypothesis was tested by attempting two different Multi-dimensional Scaling models, a one-dimensional model and a two-dimensional one. The two-dimensional model provided better fit for all four attributes (x^2, $p<0.001$), implying that there is significant evidence for the fact that the ratings for the positive and (inversed) negative poles of the attributes do not underlie a single latent one-dimensional construct. Table 2.2 depicts the squared multiple correlation coefficient R^2 between the latent one-dimensional construct and the original attribute. One can note that no clear differences between negation and opposition attributes emerged in this limited study (see figure 2.2). The attribute standard-creative (original term in Dutch fantasieloos-creatief) shows the least agreement between the two opposite poles.

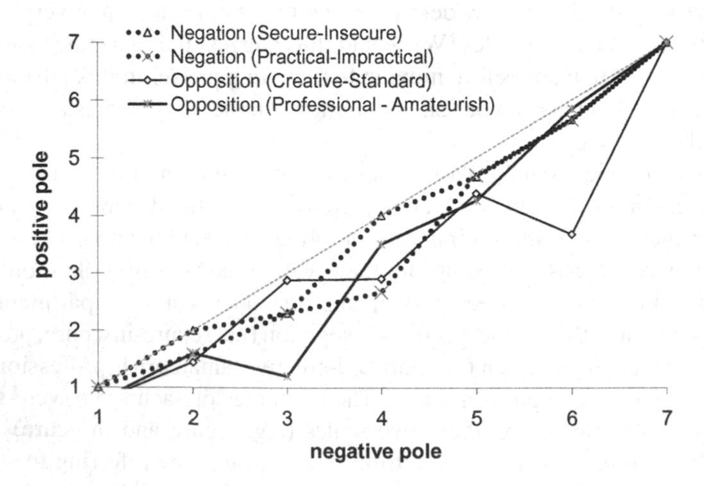

Fig. 2.2 Consistency between positive and negative poles for negation and opposition attributes. Negative scores were inverted.

2.3.3 *On the Measurement of Meaning*

Due to the notion of bipolarity, semantic differential scales (Osgood et al., 1957) are traditionally employed in RGT rating practices. In (Karapanos and Martens, 2008) we argued for comparison scales (such as paired comparisons) as an alternative, since such scales are known to be less sensitive to contextual effects than single-stimulus scales, such as the semantic differentials. de Ridder (1996) explored this issue in an experiment where quality assessments were made on a set of images. The same set of images was embedded in a larger set of images with a positively or a negatively skewed distribution in quality. The sensitivity of different scales (such as single-stimulus and comparison scales) to these contextual effects was assessed. It was found that contextual effects were negligible only when comparison scales were used. Thus, in the context of the RGT, contextual effects such as individual differences in prior experiences or order effects during the rating process will evidently make individuals' judgments less reliable when single-stimulus scales are employed.

2.4 Analyzing Personal Attribute Judgments - An Initial Exploration

One of the advantages of personal attribute elicitation techniques such as the Repertory Grid, i.e. their ability to provide rich insights into individual's perceptual and cognitive processes, is also one of their dominant limitations as this complicates the analysis phase.

Personal attribute elicitation techniques emphasize the idiosyncratic nature of perception and evaluation of objects. In other words, individuals perceive interactive products through different, individual "templates". This in turn leads to diversity in the obtained attributes and attribute ratings. Some people may use entirely different attributes to evaluate a product, whereas others may use the same attributes but apply them differently. Researchers are then confronted with as many idiosyncratic views as participants. In practice, the consequence is either an idiosyncratic analysis with a "narrative" summarization (e.g. Hassenzahl and Trautmann, 2001) or the use of average models.

Idiosyncratic approaches often discard the quantitative structure of elicited attributes and utilize only their semantic information. The researcher attempts to summarize the semantically distinct attributes and to qualitatively identify relations among them. Techniques such as laddering and pyramiding, which probe the participant to further elaborate on the elicited attribute, provide rich qualitative insight into the elicited attribute. However, as these approaches do not properly utilize the quantitative structure of attributes, they render the Repertory Grid as yet another qualitative technique, and thus undermine its true value which is in quantifying rich qualitative insights.

Quantitative analysis procedures typically rely on exploratory multivariate techniques such as Principal Components Analysis (PCA) and Multi-Dimensional

Scaling (MDS). These techniques aim at modeling relations between stimuli (e.g. interactive products), attributes (e.g. "professional - unprofessional") and overall judgments (e.g. preference). More specifically, MDS looks for a K-dimensional configuration for the stimuli such that the coordinates of the stimuli in the configuration space along different axes can be monotonically related to the observed attribute ratings of the participants (Martens, 2003).

Figure 2.3 illustrates a two-dimensional MDS configuration with two stimuli and two attributes. The relative positions of the stimuli on a given attribute axis reflect participants' ratings for the stimuli on this attribute. For instance, website *j* can be perceived as being both more legible and colorful than websites *i* and *k*.

An important motivation for MDS is the principle of homogeneity of perception which states that attribute judgments from different participants are related and thus can be represented in a common configuration space (Green et al., 1989; Martens, 2003). This view, although it often holds in perceptual judgments, has recently been challenged in more cognitive judgments where the quality dimensions of interactive products are assessed.

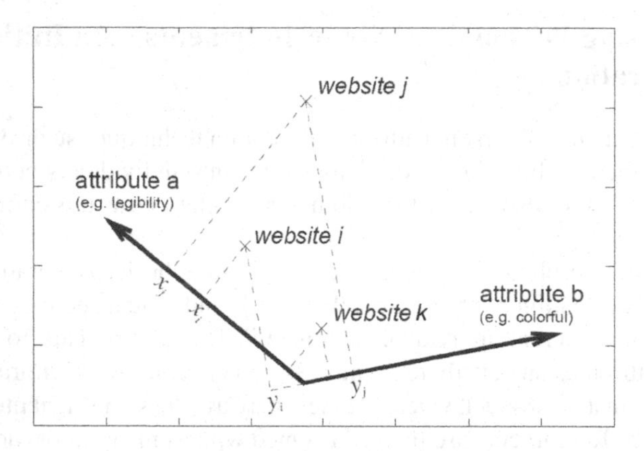

Fig. 2.3 A two-dimensional MDS configuration of three websites using ratings from two attributes. Website *j* is perceived as more legible and colorful than websites *i* and *k*.

To our knowledge, all RGT approaches up to date have been employing an average two-dimensional configuration for the quantitative analysis of personal attribute judgments from all participants. As it will become evident in Chapter 3, such averaging analysis procedures in low dimensional spaces, which constitute the common practice in the analysis of Repertory Grid data, fail in providing insight into the majority of elicited attributes.

This chapter proposes an initial Multi-Dimensional Scaling procedure that aims at inquiring into the diverse views that different individuals might have on a set of products. The procedure consists of two steps: a) identifying homogeneous groups of individuals, and b) eliciting MDS configurations for each homogeneous group to understand how they perceive and evaluate the products being studied.

2.5 The Study

We applied this procedure in a study that aimed to identify the differences, if any, between users' and designers' perceptions of products. Previous research has emphasized that designers often fail in accurately predicting users' preferences (Kujala and Kauppinen, 2004). Design decisions may thus be largely impacted by the implicit values shared within a design team. Den Ouden (2006) noted that most reasons for product returns are attributed to decisions made early in the conceptual phase of design, while it has been observed that often designers are unable to trace back decisions made in earlier phases of development (Koca et al., 2009). In this study we attempt to inquire into users' and designers' views on conceptual designs.

2.5.1 Method

2.5.1.1 Participants

Eleven "designers" and eleven potential end-users participated in the study. Designers were employees of the R&D department of an international company developing document systems. They were all involved in the conception and realization of *TouchToPrint*, which is a new way of personalizing the use of a shared printer by means of fingerprint recognition. They ranged from usability experts and market experts to software engineers and documentation experts. We refer to them as 'designers' since they were all stakeholders in the conceptual design phase. Users were researchers and employees from our department who had no prior knowledge of the product under development.

2.5.1.2 Stimuli

The *TouchToPrint* concept and five alternative proposals for interacting with a shared printer were selected for the triading process. These were the *Touch & Select* concept, which was identical to TouchToPrint but also offering modification possibilities at the printer (e.g. final selection of documents, stapling); The *Badge* concept where user identification takes place by holding an identity badge close to a sensor; the *Scroll List* concept where user identification takes place by scrolling through a list of usernames; The *Pin Code* concept where the user types her personal pin code to identify herself to the printer; and the *Direct Print* concept where no user identification is required, which reflects the most common interaction today. All concepts were presented in the form of storyboards describing a usage scenario of the relevant concept and highlighting the specific details of the relevant concept (see figure 2.4).

2.5.1.3 Procedure

The six storyboards were combined in three *triads*, i.e. combinations of three elements. These three triads were expected to elicit user perceptions related to three

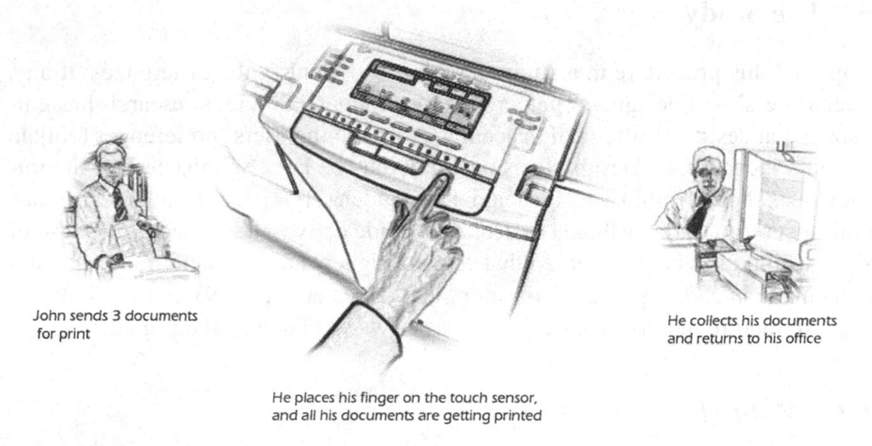

John sends 3 documents
for print
He places his finger on the touch sensor,
and all his documents are getting printed
He collects his documents
and returns to his office

Fig. 2.4 An example storyboard describing one of the concepts used in the study.

layers of interaction with the printer: the option of user identification at the printer (Direct print - Scroll list - Pin code), different user identification mechanisms (Scroll List - TouchToPrint - Badge), and the option of document selection at the printer (TouchToPrint - Direct print - Touch & Select). The order in which the triads were presented was counterbalanced between participants.

For every triad, participants were asked to *"think of a property or quality that makes two of the products alike and discriminates them from the third"*. Designers were instructed to think of ways that would be meaningful to users. Users were defined as employees at a university department. This process was repeated until a maximum of six attributes were elicited for each triad.

After attribute elicitation, participants were asked to remove duplicate attributes and rank the remaining attributes according to their importance. Finally, participants were asked to rate all products on their personal attributes, as well as on *preference* and *dissimilarity*. In contrast to the traditional Repertory Grid approach, we employed paired comparisons instead of semantic differentials, as this was a priori expected to deliver more stable results (de Ridder, 1996). While in the semantic differential technique only one product is being rated and thus being compared to an implicit reference point, in paired comparison two products are being compared on a specific attribute. Out of the six products one can form up to $n(n-1)/2 = 15$ pairs. To reduce the number of pairs we employed a balanced incomplete design (Sandt Van de, 1970; Furlan and Corradetti, 2006) with 9 total pairs and every of the six products participating in 3 pairs. According to this design, the total number of pairs can be calculated from equation 2.1:

$$TotalNumberOfPairs = \frac{n\lambda}{2} \tag{2.1}$$

where n is the number products and λ the number of pairs in which we want each product to participate in.

2.6 Analysis Procedure

The analysis procedure consists of three steps. First, a user segmentation map that expresses the diversity among individuals is derived from the collected dissimilarity ratings by means of Multi-Dimensional Scaling (MDS) (Martens, 2003). Homogeneous groups of users are identified within this map by means of (hierarchical) clustering. Secondly, attributes are classified into categories based on their semantic content. This semantic classification will then be contrasted to attribute ratings. Third, perceptual maps are created from the attribute, dissimilarity and preference ratings to express how homogeneous groups of participants perceive the products being studied.

2.6.1 Identifying Homogeneous User Groups in the User Segmentation Map

In this step a user segmentation map that expresses the diversity among users is derived from their dissimilarity ratings by means of Multi-Dimensional Scaling. To create the user segmentation map, we define the distance $D_{i,j}$ between participants i and j based on the correlation $R_{i,j}$ between their dissimilarity scores. Derived distances are then visualized in two or more dimensions using the MDS tool XGms (Martens, 2003). Figure 2.5 displays a two dimensional configuration of designers and users. The closer two individuals are in the two-dimensional space, the more their ratings of overall dissimilarity of products correlate. The dimensionality of the configuration may be judged by the experimenter using the Stress Value which is an index of goodness of fit of the model. In this case, the two dimensional visualization was judged as adequate (stress value S=0.18) (Clarke, 1993).

$$D_{i,j} = 1 - R_{i,j}^2 \tag{2.2}$$

$$R_{i,j} = \frac{\Sigma_k D_i(k) D_j k}{\sqrt{D_i{}^2(k) D_j{}^2(k)}} \tag{2.3}$$

Hierarchical clustering performed on the reduced two-dimensional space (with minimum variance) reveals two main clusters that can be further subdivided into five more or less homogeneous participant groups. Groups 3 and 4 consist entirely of end users, while groups 1, 2 and 5 consist mostly of designers. Identification of the designers reveals that group 1 consists mostly of technically-oriented designers, while group 2 consists mostly of user-oriented designers.

2.6.2 Classifying Attributes for Interpersonal Analysis

In this step, attributes are submitted to a content analysis (Hsieh and Shannon, 2005) where key concepts, i.e. distinct categories of attributes, emerge from the data and attributes are subsequently classified into one of the elicited categories. A total of

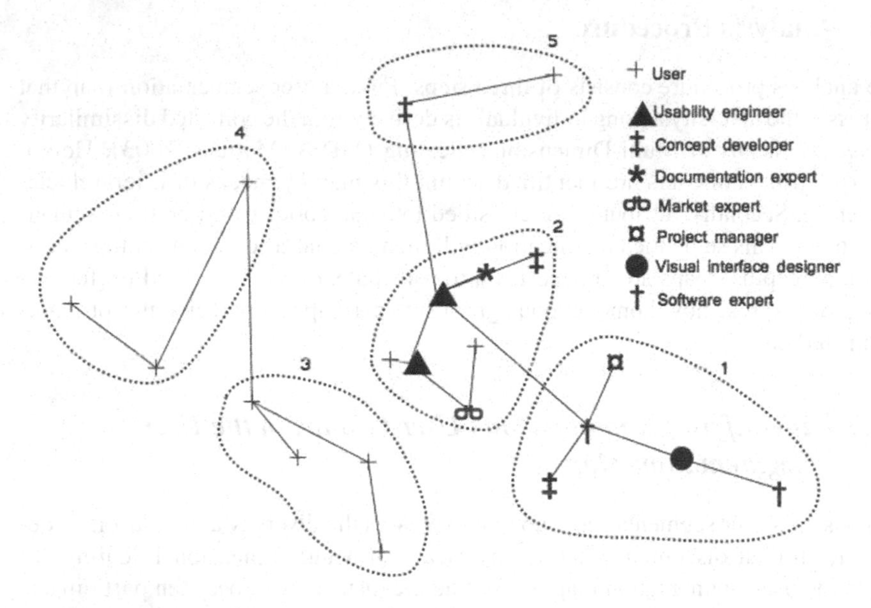

Fig. 2.5 User segmentation map based on correlations between dissimilarity scores

81 attributes for designers and 95 for users were obtained in the study (6 to 11 per participant). Two rounds of analysis were performed: an exploratory semantic classification of attributes, followed by a confirmatory analysis of the classification. For the first round, sixteen semantically distinct attribute categories were first formed out of the data. To minimize the researcher's bias, the naming of the attribute categories was restricted to choosing one of the attribute names that reflect this semantic value. Subsequently, the first author and two additional experimenters independently assigned every attribute to one of the sixteen categories (Table 2.3). Interrater agreement (Fleiss et al., 2003) of the initial classification was satisfactory (K=0.72). All sixteen categories were then classified into three overall classes: *Effectiveness*, *Efficiency*, and *Emotional Appreciation* (interrater agreement, K= 0.80).

During the confirmatory analysis of the classification, statistical consistency across attributes within the same category was being sought. Attribute scores were submitted to a cluster algorithm where Euclidean distances between attributes were calculated and visualized in two or three dimensions by means of Multi-Dimensional Scaling. Outlier attributes, i.e. ones that did not belong in the same or a neighbor cluster to the one that were characterized primarily by attributes of a given semantic category, were identified. The prospect of transferring the outlier attribute to one if its statistically-neighbor categories was explored. If there was no argument for a transfer to another category, the attribute was deleted.

Table 2.3 Classification of attributes elicited in the study into 16 categories and three broad themes

Effectiveness (32%)	**Efficiency (51%)**	**Emotional Appreciation (17%)**
1. Secure	5. Fast	12. Personal
2. Reliable	6. Error-prone	13. Modern
3. In control	7. Brainless	14. Privacy
4. Multi-user	8. Cognitive load	15. Hygienic
	9. Effort in searching documents	16. Environmental friendly
	10. Fast (total time)	
	11. Easy-to-learn	

2.6.3 Charting Perceptual Maps for Homogeneous Groups of Users

In this step, a perceptual map is created for each homogeneous group of users. First, a configuration of the stimuli is established in a k-dimensional space where their distance represents their degree of dissimilarity. Next, attribute and preference vectors are fitted into these spaces by performing multiple regressions between the stimulus configuration (as independent variables) and the attribute scores (as the dependent variables).

Figure 2.6 illustrates the perceptual maps of the five groups. All five perceptual spaces were adequately visualized in two dimensions (χ^2 test on goodness of fit, $p<0.001$). Only significant attributes were retained in the final visualization, i.e. ones that their confidence ellipse did not include the zero point of the configuration. Attributes are represented by the bold vectors and preferences by the light vectors.

The interpretation of the perceptual maps can be done in three steps: a) identifying the similarities and dissimilarities between products, b) interpreting product dissimilarities by employing the attribute vectors, and c) inferring the values of group of participants by relating individual preferences with perceived attributes.

2.6.3.1 Identifying Dissimilarities between Products

One can note that all perceptual spaces are clustered in two regions, one containing *TouchToPrint*, *Badge* and *Touch & Select* and the other containing *Scroll List* and *Direct Print.* The perception of Pin Code in comparison to the other systems is less consistent across different perceptual spaces. It is further apparent that the configuration of products within these regions differs across the perceptual spaces. Groups 1 and 4 perceive differences between Scroll List and Direct Print while Groups 2, 3 and 5 perceive these two products as relatively similar. Further, Groups 1 and 5 perceive TouchToPrint and Badge as very similar in comparison to Touch & Select while Groups 2, 3 and 4 perceive relatively strong differences between TouchToPrint and Badge.

2.6.3.2 Interpreting Product Dissimilarities

As mentioned all perceptual spaces show two main clusters of products. One may notice a strong preference for products in the first cluster (i.e. only two participants prefer products from the second cluster). This can be further explained by analyzing the directions and the type of attributes. It is evident that most attributes point to product cluster 1 while only three attributes point to cluster 2. A close look at the content of these three attributes provides some more insight. Two out of the three attributes are essentially the contrast poles to a negative attribute of the comparison products. This is also evident when looking into the element differentiators. Participants indicated that "TouchToPrint is not reliable" and that "it is privacy threatening", thus other products will score better on reliability and privacy. It seemed that participants could more easily identify qualities in products of cluster 1 that were less familiar to them than in the traditional products from cluster 2.

Within product cluster 2 one can notice differences among groups. Groups 1 and 4 perceive differences between Scroll List and Direct Print while Groups 2, 3 and 5 perceive these two products as relatively similar. Group 1 seems to perceive differences between Scroll List and Direct Print in two attributes: control and personal. Apparently, for Group 1, Scroll List provides more control in the interaction as documents are only printed when the user is identifying himself to the printer. Moreover, it is more personal than Direct Print as every user has a personalized access to the printer. More interesting though is the observation that Group 3 does not regard scroll list as personal despite having this personalization feature. Apparently, users from Group 3 distinguish between the personalization features of Scroll list and those of TouchToPrint, Badge and Touch & Select. One explanation could be that the perception of personalization for Group 3 is influenced by the perception of security (i.e. with scroll list one can identify himself as another person). Further, group 4 seems to perceive differences between Scroll List and Direct Print in two attributes: environmental friendly and fast. One can notice that group 4 differs from other groups in the perception of the attribute fast. For group 4 the two extremes are Scroll List and Direct Print where the second is fast as it involves no interaction with the product and the first involves tedious interaction. It is also evident from preferences that a product quality for this group is a minimum interaction with the product. The other groups however perceive that direct print requires additional effort for identifying the documents among the pile of all printed ones, thus Direct Print is not fast. It is evident that the attribute fast is not statistically consistent between groups but it is consistent within groups. This demonstrates the value of the two-step classification process as it enables us to gain insight in the different perceptions of an indistinguishable product quality.

Further, within cluster 1 one can notice differences in the relation between Badge and TouchToPrint as compared to Touch and Select. Groups 1 and 5 perceive TouchToPrint and Badge as very similar to Touch & Select while Groups 2, 3 and

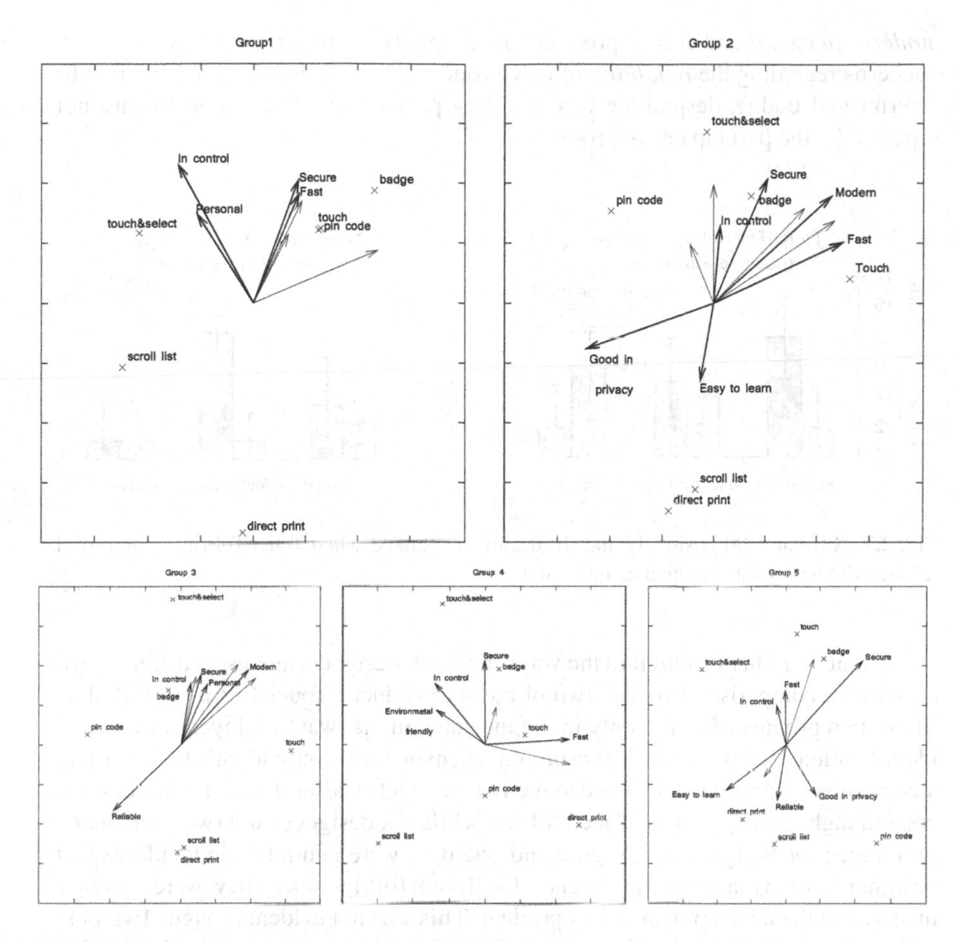

Fig. 2.6 Perceptual maps for groups 1 to 5. The light vectors represent individual preferences, while the other vectors represent the most significant attributes.

4 perceive relatively strong differences also between TouchToPrint and Badge. All three groups think that Badge provides more control in the interaction than TouchToPrint. Further, Group 4 perceives Badge as more secure than TouchToPrint while Group 2 perceives a tradeoff in TouchToPrint which is the capturing of personal biometric data (i.e. privacy) and Group 3 has concerns about the reliability of TouchToPrint due to the novelty of the technology.

2.6.3.3 Inferring Users' Values

By relating perceived attributes to individual preferences one can infer priorities existing in users' values as indicated within this limited set of products. One can note that for individuals in group 1, *secure* and *fast* are the most important product qualities influencing their preference judgment, while participants in group 3 value *modern*, *personal* and *secure* products. Participants within group 3 have negative concerns regarding the *reliability* of new products, such as Touch & Select, TouchToPrint and Badge, despite the fact that they prefer them. Such concerns are not reported by the participants in group 1.

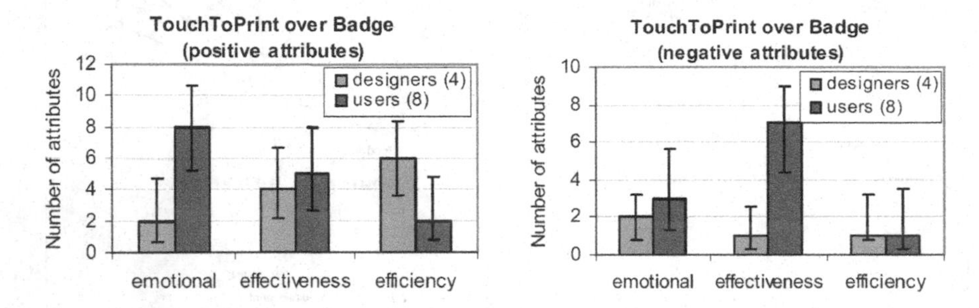

Fig. 2.7 Attributes (a) positively and (b) negatively ranked when TouchToPrint is preferred (along with 95% exact confidence intervals)

To acquire richer insight into the ways in which designers and users differ we focused on a comparison between two of the six products: TouchToPrint and Badge. These two products differed only in the mechanism that was employed for the user identification process: a touch sensor and a sensor for an individual's badge. First, a comparison of preference between the two products was performed. Four designers and eight users preferred TouchToPrint while six designers and two users had a preference for Badge. One designer and one user were neutral. One would expect designers to have a strong preference for TouchToPrint since they were recently involved in the development of this product. This was not evident though. Two possible explanations might be given: a) Badge was not yet implemented by the specific company, therefore treated as future development by designers and thus assigning to it greater value than to the already existing TouchToPrint or b) potential drawbacks of TouchToPrint were only evident after use, and hence more obvious to the designers who had actually experienced the product. The users, who had no actual experience with TouchToPrint, seemed to value it higher than designers did.

To further understand this discrepancy between designers' and users' preferences, we analyzed their perceptions for these two products. Figure 2.7a illustrates the reasons supporting preference for TouchToPrint over Badge, as it shows the number of attributes that are positively ranked when TouchToPrint is preferred. While users' most frequent reason for preference for TouchToPrint was emotional attributes, for designers it was efficiency attributes. All attributes in the effectiveness category were related to security. TouchToPrint was perceived as more secure than Badge, both by designers and users. Users' most frequent negative concerns, shown in Figure 2.7b, were related to reliability (5 out of the 7 effectiveness attributes had to do with reliability). This is also evident in Figure 2.8 where we can observe that only two designers expressed reliability concerns and ranked them as the 6th and

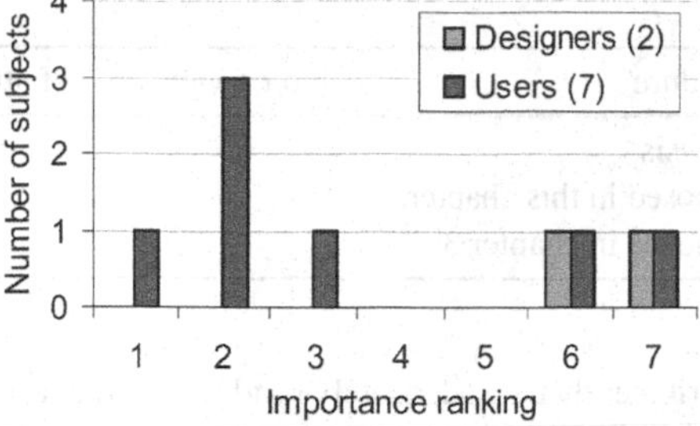

Fig. 2.8 Importance rankings for reliability attributes for designers and users. Only 2 designers reported reliability concerns in contrast to 7 users. Five users ranked reliability attributes within their three most important attributes.

7th most important attributes while five users ranked reliability within their three most important concerns. Hence, although most users prefer TouchToPrint, they have some concerns that can potentially turn into failures to motivate use.

2.7 Discussion

In this chapter a simple and effective technique was presented for acquiring diverse views on Repertory Grid data. The analysis revealed systematic differences between the product qualities that users and designers appreciated within this limited set of products. The proposed analysis procedure, however, relies on a number of assumptions.

First, it assumes the individual as a homogeneous entity. Attribute judgments of a single individual are assumed to be well predicted by a single two-dimensional configuration, and the emphasis lies in identifying individuals in a sample that form an approximately homogeneous user group. At certain occasions this might be a desired property of the technique, e.g. when the researcher wants to characterize single individuals such as identifying differences in the perceptions of different stakeholders in a design team. Diversity could however exist not only across different individuals, but also within a single individual, in the sense that different attribute judgments of a participant may reveal different, complementary, views.

Second, diversity across individuals was explored using a single measure, that of overall dissimilarity between the products. This approach assumes that individuals are able to consistently summarize all attribute judgments in a single rating of dissimilarity, an assumption that is ubiquitous in the field of psychometrics were dissimilarity serves to summarize various perceptual judgments such as the perceived blur and noise in an image (Martens, 2003). One might, however, question the degree to which this holds in more complex settings where various cognitive judgments are made for the quality of interactive products.

Table 2.4 Number (and percentage) of attributes being adequately modeled by the resulting views of the averaging analysis and the two proposed procedures.

Analysis Procedure	No of views	No of attributes (%)
Averaging analysis	1	65 (43 %)
Procedure proposed in this chapter	5	62 (41 %)
Procedure proposed in chapter 3	2	85 (56 %)

Table 2.5 Number of attributes that are adequately modeled by the resulting one or two views for each participant.

Subj.	Description	view a	view b	remain
1	User 1	6	-	1
2	User 2	10	-	1
3	User 3	-	-	13
4	User 4	6	4	1
5	User 5	-	-	10
6	User 6	6	-	2
7	User 7	9	-	3
8	User 8	6	-	4
9	User 9	10	-	-
10	User 10	9	-	1
11	User 11	6	-	3
12	Concept Developer 1	11	-	1
13	Concept Developer 2	6	-	4
14	Concept Developer 3	6	-	1
15	Documentation Expert	7	-	1
16	Interface Designer	-	-	2
17	Market Expert	8	-	1
18	Project Manager	7	-	-
19	Software Expert 1	6	-	1
20	Software Expert 2	5	-	2
21	Usability Expert 1	-	-	9
22	Usability Expert 2	6	4	-

Third, the technique did not explicitly optimize the goodness of fit of the diverse models for the analyzed attributes. As a result, while the technique succeeded in identifying the differences between users and designers in terms of overall dissimilarity, it might fail in accounting for more attributes than the traditional *"averaging"* analysis. In the next chapter, a new procedure is described that explicitly aims at increasing the number of attributes that are adequately modeled based on two criteria introduced below. In this procedure, two goodness-of-fit criteria are defined for assessing whether or not an attribute is adequately predicted by a given model: a) the amount of variance R^2 in the attribute ratings accounted for by a given model, i.e.

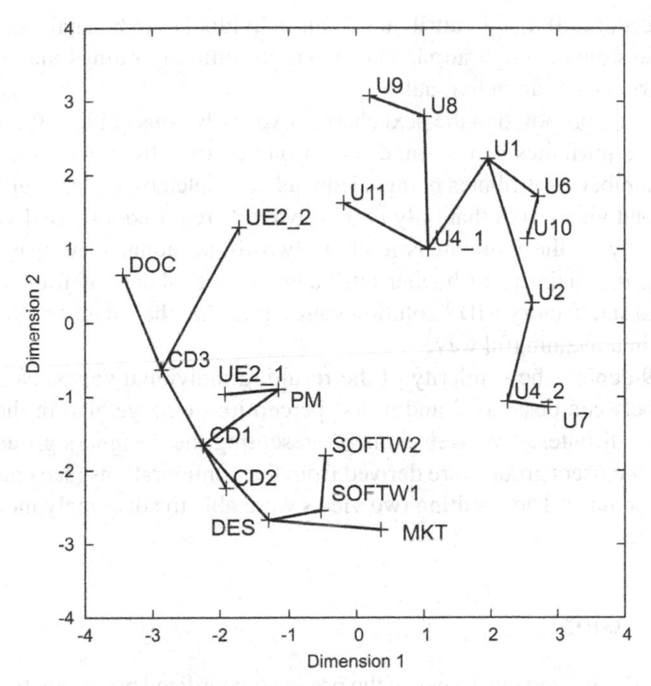

Fig. 2.9 Two dimensional visualization of the dissimilarity of the individual views derived from the reanalysis of the data using the procedure proposed in chapter 3. Views connected through lines belong to the same cluster. Abbreviations used: Doc=Documentation Expert, CD=Concept Developer, UE=Usability Expert, PM=Product Manager, SOFTW=Software Expert, DES=Visual Interface Designer, MKT=Market Expert, U=User

MDS configuration, and b) the ratio of the maximum range of the predicted scores for an attribute divided by the standard deviation σ_k of the estimation error in the attribute scores.

In the remaining of this section, we employ these criteria for comparing the three different analysis procedures: a) the traditional averaging analysis, b) the analysis proposed in this chapter, and c) the second analysis procedure proposed in chapter 3 that aims at optimizing the goodness of fit for the majority of the attributes.

Table 2.4 depicts the number (and percentage) of attributes being adequately modeled by the resulting views of the three different analysis procedures. The averaging analysis resulted in a total of 65 or 43 % of all attributes being adequately modeled by a single two-dimensional configuration. Surprisingly, the analysis procedure proposed in this chapter, despite resulting in five diverse views, performed worse than the averaging analysis when we are concerned about the number of attributes being adequately modeled by the resulting views. Even when analyzing together the attributes of individuals that agreed on the overall dissimilarity of the stimuli, a wealth of attributes still remained inadequately modeled by the resulting shared configuration. This challenges the validity of two assumptions made in this analysis procedure: a) that all attributes of an individual may be analyzed by a single two-dimensional configuration, and b) that dissimilarity ratings may be used to model all attributes of an individual.

The procedure proposed in the next chapter explicitly aims at identifying whether an individual's attributes can be modeled in one or two diverse views. Table 2.5 depicts the number of attributes being adequately modeled by the first and (in some cases) a second view. Note that only two individuals required a second view while for the majority of the individuals a single two dimensional view was sufficient for modeling the majority of his/her attributes. The attributes of four individuals resulted in no satisfactory MDS solution suggesting that their data are too noisy to be analyzed in a meaningful way.

Figure 2.9 depicts the similarity of the resulting individual views. Note that the differences between designers' and users' perceptions emerge also in the analysis of individual attributes. Two views, one representing the designers group and one representing the users group were derived from the configurations (see chapter 3 for the exact procedure). The resulting two views were able to adequately model a total 85 attributes (56%).

2.8 Conclusion

This chapter highlighted two issues in the use of standardized psychometric scales in measuring users' experiences with products. Firstly, a-priori defined measurement scales are inherently limited in missing potentially relevant measures that the researchers did not consider as relevant to a given context. Secondly, in certain cases, rating products on measures defined by the researchers is not always a meaningful activity for the user, for example when the user does not consider a quality dimension as relevant for the specific product or is unable to situate the question in her own context.

We introduced personal attribute judgments and respective attribute elicitation techniques as an alternative to standardized psychometric scales. These approaches have the potential to account for the richness of user experience and bridge the gap between formative and summative evaluation by quantifying rich qualitative insight. We highlighted one technique, namely the Repertory Grid Technique which has recently become popular in the field of Human-Computer Interaction and User Experience.

We noted the lack of appropriate statistical analysis techniques that lead to practices that either treat RGT as yet another qualitative technique, or employ averaging models which undermine the core motivation of RGT and relevant techniques. We argued that the true value of RGT is in quantifying rich qualitative insights.

We identified the two dominant approaches in analyzing repertory grid data: a) a qualitative, idiosyncratic approach with a narrative summarization, and b) a quantitative approach that employs averaging procedures using exploratory multivariate techniques. We argued that these two approaches are limited in two respects. Qualitative approaches do not take properly into account the elicited attribute ratings and thus do not fully exploit the true value of the Repertory Grid Technique, which is to quantify rich qualitative insights. Averaging procedures, on the other hand, treat diversity among participants as error and thereby contradict the basic idea of The Repertory Grid and relevant personal attribute elicitation techniques.

Last, we proposed an initial quantitative technique that aims at inquiring into the diverse views that different individuals might have on a set of products. The technique employs Multi-Dimensional Scaling in a) identifying homogeneous groups of individuals, and b) eliciting a different view for each homogeneous group of individuals.

The technique was applied in a study that tried to assess the differences between designers' and users' perceptions on a set of early conceptual designs. The results from the study corroborated prior findings (Kujala and Kauppinen, 2004) suggesting that designers may fall short in accounting for users' views on their product and foreseeing their preferences. Surprisingly, the diversity between the designers and users was larger than the diversity across different user groups. This might be affected by the different exposure of these two diverse groups to the different concepts, thus leading to a relative agreement across the different users' in comparison to the designers group. One has to note that designers were explicitly asked to reflect on the way that this specific user group would perceive this set of products. In this way the gap between designers' and users' perceptions was minimized to reflect actual design practice where assumptions and claims are made for a specific user group. These insights highlight the need for grounding design decisions that are made early in the concept design phase on quantifiable empirical data about users' preferences.

Chapter 3
Analyzing Personal Attribute Judgments

Abstract. This chapter presents a second Multi-Dimensional Scaling procedure that aims at identifying diverse views even within single individuals. The technique is applied on an existing dataset (Heidecker and Hassenzahl, 2007). It is illustrated that the - traditional - averaging analysis provides insight to only 1/6th of the total number of attributes in the example dataset. The proposed approach accounts for more than double the information obtained from the average model, and provides richer and semantically more diverse views on the set of stimuli.

3.1 Introduction

In the previous chapter a simple and effective technique was presented for acquiring diverse views on Repertory Grid data. This technique, however, relies on a number of assumptions. First, it assumes that all attributes of an individual may be analyzed by a single 2D view, and second, that individuals are able to consistently summarize all attribute judgments in a single judgement of overall dissimilarity. As such, it discards diversity existing within a single individual and does not explicitly optimize the goodness of fit of the diverse models for the analyzed attributes. It was shown that while the technique succeeded in identifying the differences in individual's perceptions in terms of overall dissimilarity, it failed in accounting for more attributes than the traditional *"averaging"* analysis.

This chapter suggests a quantitative, exploratory Multi-Dimensional Scaling procedure to account for the diverse views that one or more individuals may have on a set of products. It will be demonstrated that even single individuals can handle more than one view on a set of stimuli. It will be shown that by averaging interesting views are overlooked due to majorization bias. The insights strongly advocate the view that the analysis of quality judgments of interactive products should not stop on a group level, but must be extended to the relations between the attribute judgments within an individual. The Repertory Grid combined with the suggested technique to analyze the resulting quantitative data is an important step towards the adequate account of homogeneity and especially diversity in individual quality judgments.

E. Karapanos: Modeling Users' Experiences with Interact. Syst., SCI 436, pp. 41–56.
springerlink.com

3.2 The Study

The data for the present analysis was taken from a study of Heidecker and Hassenzahl (2007) of individuals' perceptions of eight university websites (figure 3.1). The study was part of a larger project aiming at understanding how the Technical University of Darmstadt (TUD) is perceived in comparison to other regional competitors. Ten individuals, all students at TUD, participated in the study.

Table 3.1 Attribute categories and examples

Attribute category	Example
Layout	Graphical layout - Textual layout
	Colorful - Pale colors
	Professional - playful
University Image	Technical studies - Social studies
	Emphasis on achievement - Average univ.
	Refers to student life - Modern organization
Information Access	Fast access to information - time-intensive
	Legible - Tangled

The eight university websites were presented to participants in the form of colored A4 screenshots of the main page. Using the Repertory Grid Technique, a number of attributes on which the eight websites differ, were elicited from each participant. Participants were then asked to rate the websites on their own elicited attributes, using 7-point Semantic Differential scales. The resulting data set consisted of a total of 118 attributes (10 to 14 per participant) on which ratings for the eight different stimuli were elicited.

3.3 A Multi-dimensional Scaling Approach to Account for Diversity

The starting point of the proposed approach is that of *identifying the different views* that each participant has on the set of products. In this step, an average model is attempted for each participant. However, attributes that are not adequately predicted by the average model (see Table 3.2) are removed and used in deriving a second model, i.e. a second view for the specific participant (Figure 3.2 illustrates two diverse views derived for one participant).

Once the diverse views of all individuals have been identified, *the similarity among them is assessed* and views are clustered into groups of increased homogeneity.

A final set of diverse configurations is formed by grouping the similar views, which are then used to model the attributes from all participants.

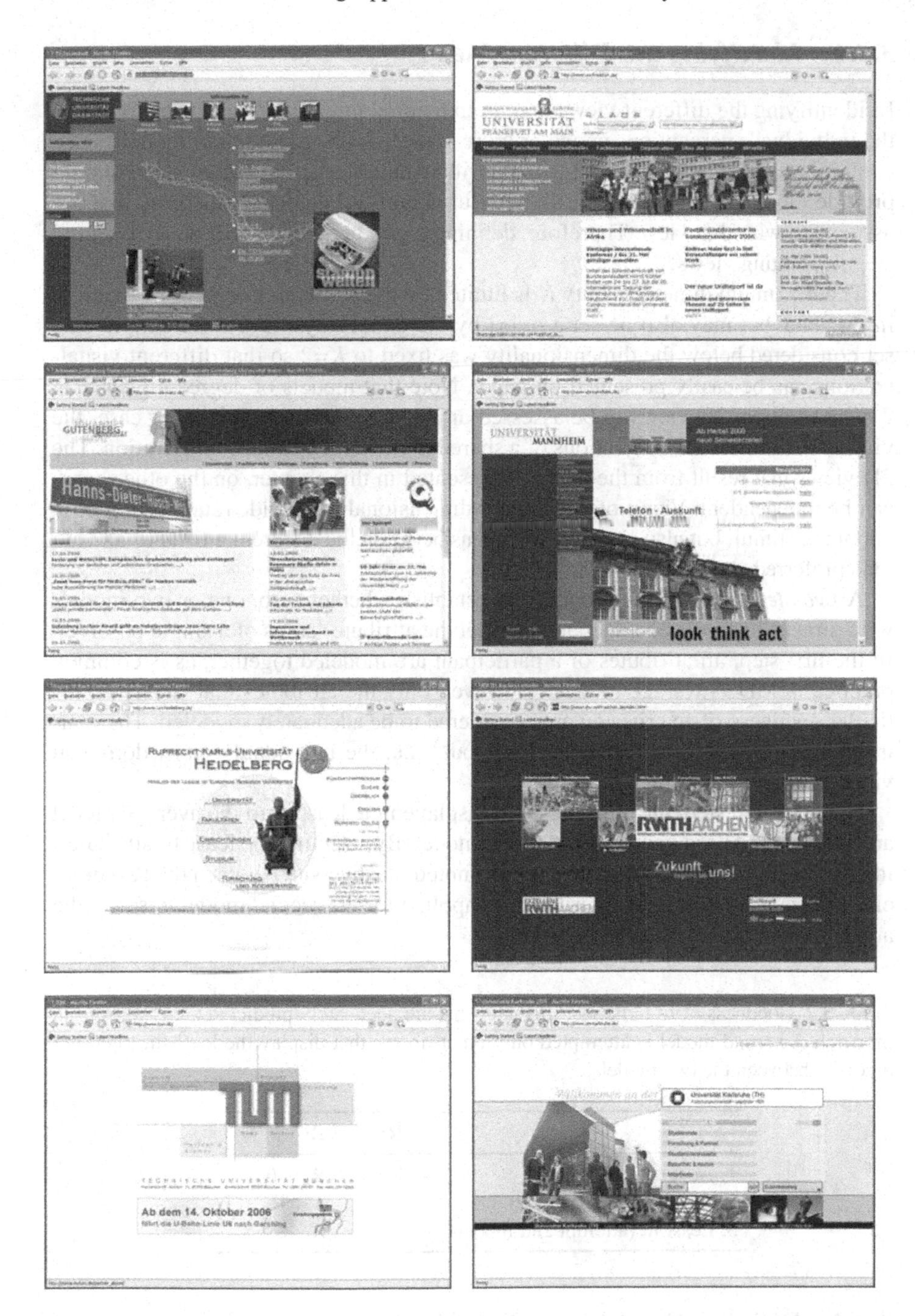

Fig. 3.1 The eight stimuli used in the study

3.3.1 Identifying the Different Views

In identifying the different views that an individual might hold, one tries to model the individual's perceptions in one or more *non-trivial K-dimensional* models, each explaining *adequately* a part of his/her attribute judgments. Individual views should provide close matches between the measured and the modeled attributes that are associated with that view. Therefore, defining a goodness-of-fit will be an essential step in creating views.

The maximum dimensionality K is limited by the number of degrees of freedom in the data, but may also be set a priori by the data analyst. For the example data set considered below the dimensionality was fixed to K=2 so that different visualizations can be easily presented on paper. Note that models of degree higher than 2 need multiple 2D views to be assessed anyhow. However, in this latter case, the views are different 2D projections of a shared multi-dimensional configuration. The 2D views that result from the analysis presented in this chapter, on the other hand, can be independent. Views of 2 or higher dimensionality provide relations not only about the stimuli but also about the relations between the different attributes and are thus preferred over 1D views.

A *two-step procedure* is proposed to establish whether zero, one or two models with dimension K=2 can adequately model the attribute scores of a single observer. In the first step, all attributes of a participant are modeled together, as is common practice in MDS (average model). However, only the attributes that satisfy a particular goodness-of-fit criterion are considered to be adequately modeled. These attributes are analyzed to form the first model, i.e. the individual's most dominant view on the set of products.

In the second step, the attributes that displayed the least fit to the average model are grouped and used to attempt a second model. By selecting the least-fit attributes, instead of all remaining attributes, we promoted the diversity between the two models. The same goodness-of-fit criteria are applied for the second model to select the attributes that are retained.

Table 3.2 Goodness of fit Criteria. Attributes that are adequately predicted are employed in model 1. A second model is attempted only on attributes that display the least fit, to ensure diversity between the two models.

	R^2	R_k
1. Adequate fit	$R^2 > .5$	$R_k > 6$
2. Average fit (Excluded)		$4 < R_k < 6$
3. Least fit (attempt 2nd model)		$R_k < 4$

3.3.2 Defining Goodness-of-Fit Criteria

We suggest a combined goodness of fit criterion. First, for an adequately predicted attribute, a substantial amount of its variance should be accounted for by the model.

This proportion of explained variance is the R^2 statistic (i.e., the squared multiple correlation coefficient). A threshold $R^2>0.5$ was set, implying that only attributes are retained for which at least 50% of their variance is accounted for by the model. A limitation of this criterion is that it is insensitive to the range of the ratings for the different stimuli on a given attribute. An attribute might make no meaningful differentiation between stimuli (e.g. if all stimuli are rated as 4 or 5 on a 7-point scale) but can nevertheless be well-predicted by a model. To account for this limitation, we combine it with a second criterion.

This second criterion is a modification of a measure originally proposed by Draper and Smith (1998). It is the ratio of the maximum range of the predicted scores for attribute k divided by the standard deviation σ_k of the estimation error in the attribute scores (1a).

$$R_k = \frac{\hat{A}_{k,max} - \hat{A}_{k,min}}{\sigma_k} \tag{3.1}$$

$$R_k{}^* = \sqrt{\frac{1}{n^2}\sum_{i,j}\frac{[\hat{A}_{ki} - \hat{A}_{kj}]^2}{k^2}} \tag{3.2}$$

A combined criterion thus takes into account both the accounted variance in the attribute as well as the range of the scores for the different stimuli (i.e. the attribute's strength). The obvious limitation of the second measure is its sensitivity to outlier scores. However, in single-stimulus scales such as the semantic differential scales, these outlier scores may actually be very valuable, since they point at the stimuli that most strongly influence the existence of the attribute scale in the first place. When using more sensitive scales such as paired comparisons (Karapanos and Martens, 2007), one might consider adopting the modified measure (3.2) that averages across differences in predictions. Draper and Smith (1998) proposed a minimum ratio value of four, meaning that any attribute predictor with a ratio below four hardly makes any distinction between the stimuli and is pretty useless. Predictors with a ratio value above ten are considered to be excellent. We decided to use an *acceptable ratio* of six for the data analysis reported in Table 3.3.

3.3.3 Two Diverse Views for One Participant

Table 3.3 illustrates the analysis process on the attribute judgments of a single participant. A first (average) model was attempted on all attributes of the participant. Attributes (2,4,6,8,12,13; in bold) were adequately predicted by the average model, using the two criteria that were discussed before, i.e. $R^2>.5$ & $R_k>6$. Model 1 was then derived by optimizing the average model only for the attributes that were adequately predicted by the average model.

Note that the R^2 and R_k values are identical (at least in this exact decimal point) for Model 1 and the average model. This implies that when removing the attributes that are not adequately predicted (even with these arbitrary criteria), the 2D

Table 3.3 Goodness of fit statistics for the two diverse models of participant one. Attributes (2,4,6,8,12,13) were adequately predicted (R^2>.5 & R_k>6) by model 1. Attributes (1,5,7,9,10,11) displayed the least fit (R_k<4) and were used to derive a second model. Attributes (5,7,9,10) were adequately predicted by model 2.

	Attribute	Avg. Model		Model 1		Model 2	
No	Variance (σ^2)	R^2	R_k	R^2	R_k	R^2	R_k
1	2.6	0.47	2.2	0.47	2.2	0.36	3.3
2	3.8	0.89	7.3	0.89	7.3		
3	0.6	0.73	4.1	0.73	4.1	0.56	2.6
4	1.9	0.98	18.6	0.98	18.6		
5	3.7	0.49	2.3	0.49	2.3	0.95	13.7
6	2.2	0.99	40.5	0.99	40.5		
7	1.7	0.48	2.4	0.48	2.4	0.99	39.6
8	6.3	0.93	9	0.93	9		
9	4.1	0.63	4.8	0.63	4.8	0.99	40.1
10	4.5	0.26	2.1	0.26	2.1	0.61	6.5
11	3.9	0.08	0.9	0.08	0.9		
12	1.9	0.88	6.8	0.88	6.8	0.48	2.8
13	5.6	0.99	50.4	0.99	50.4	0.85	5.5

configuration space (which is represented in the model parameters) displays virtually no change. In other words, the attributes that were removed (according to the arbitrary criteria) had no contribution to the configuration space. Thus, the information contained in these attributes is not modeled when attempting an averaging analysis and therefore it is lost.

Out of all the attributes that were not adequately predicted, attributes (1,5,7,9,10,11; in italics) displayed the least fit by model 1, i.e., R_k<4. These were used to derive a second model. Out of them, only attributes (5,7,9,10) turned out to be adequately predicted by model 2, using the same goodness of fit criteria as used in model 1.

Figure 3.2 illustrates the different insights that the two diverse views bring. One can note that the two views highlight semantically different attributes. Each attribute is visualized as an arrow, i.e. a dimension, on which the relative positions of the websites can be compared. The length of each arrow indicates the strength of the attribute, reflecting the variance in the predicted attribute ratings for the different stimuli; on some attributes all websites might be rated as 4 or 5 on a 7-point scale, while others might make strong differentiations between sites, i.e. across the whole range of the scale.

The first view provides overall three different insights. First, that the universities of Frankfurt, Manheim and Mainz are perceived as putting *less emphasis on achievement*, as compared to the remaining five universities. This may be induced by the websites but may also reflect prior beliefs of the individual. Second, the websites of the universities of Mnchen, Aachen, Karlsruhe and Heidelberg have a more *professional layout* as opposed to the remaining four which have a more *playful* one.

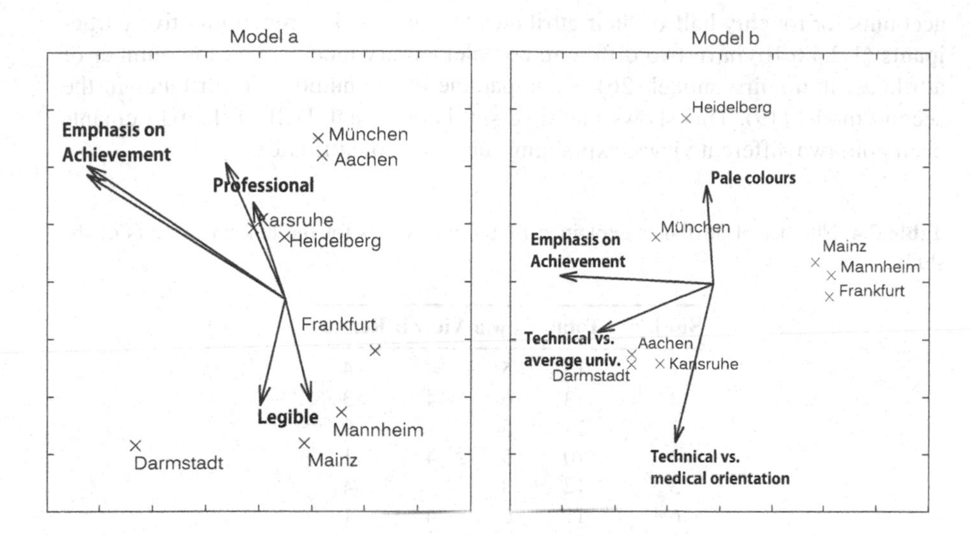

Fig. 3.2 Two diverse views for one participant

Last, the participants perceive this same group of websites as legible as opposed to the remaining four in the upper part of the figure that are perceived as having no clear structure.

The second view partly provides overlapping information (emphasis on achievement), but also gives three new insights. First, the website of the University of Heidelberg is differentiated from all others by having a less colorful layout. Second, the Universities of Darmstadt, Aachen and Karlsruhe are differentiated as universities that provide primarily technical studies, as opposed to the universities of Mainz, Mannheim and Frankfurt that are referred to as universities of average quality, and third, as opposed to the university to Heidelberg that is perceived as a university offering primarily medical studies.

Note that an attribute may range from being purely descriptive, i.e. referring to specific features (e.g. allows searching), to having an evaluative tone, e.g. referring to the perceived quality of the product (e.g. easy to use) or the product's overall appeal (e.g. good). This enables the researcher to gain a better understanding of the inferences individuals make as they form evaluative judgments of products.

3.3.3.1 The Resulting Views

Table 3.4 summarizes the results of the above analysis for all ten participants. For two of the ten participants (7, 8), no substantial agreement between their attribute judgments is observed, i.e., no satisfactory MDS-model can be derived. This implies that they either have as many different views as their attribute judgments, or more likely, that their ratings are too noisy to be analyzed in a meaningful way. For another three participants (3,5,9) only one satisfactory model is determined, which accounts for roughly half of their attributes (17 of 37). The remaining five participants (1,2,4,6,10) have two different, complementary models, i.e., the number of attributes in the first model (26) is comparable to the number of attributes in the second model (19). This shows that diversity is prevalent. Half of the participants even hold two different views, explaining subgroups of attributes.

Table 3.4 Number of attributes explained by the two views for the ten participants of the study.

Subj.	Total	View a	View b	Remain
1	13	5	4	4
2	13	6	4	3
3	14	5	-	9
4	10	5	4	1
5	12	8	-	4
6	11	6	4	1
7	13	-	-	13
8	11	-	-	11
9	11	4	-	7
10	10	4	3	3

All together, 13 different views emerged from the ten individuals. These views may partly overlap, which motivated us to group similar views and identify the major diverse of this user group.

3.3.4 Assessing the Similarity between Different Views

In grouping the diverse views one has to derive a distance measure that reflects the degree of dissimilarity between configurations. Each configuration can be regarded as a *NxK* matrix, where N is the number of stimuli and K the number of dimensions of the configuration space. The distance between configurations X_n and X_m can be calculated using the *MATFIT* procedure, developed by Ramsay (1990). *MATFIT* seeks for a transformation matrix M that minimizes the distance measure:

$$d^2 = trace[(X_n M - X_m)^t](X_n M - X_m)] \quad (3.3)$$

An arbitrary *KxK* transformation matrix M was applied. The procedure was repeated with the matrices in reverse order as a means to calculating both distances: with X_n as independent and X_m as dependent, and vice versa. The resulting distances were visualized in three dimensions using the program *XGms* (Martens, 2003). A hierarchical (minimum variance) clustering algorithm was applied to the 3-D configuration (a cluster is denoted by the lines connecting the different views). Figure 3.3 represents a 2-D perspective on the 3-D configuration of individual models. Note that the distances in this 2D perspective do not necessarily reflect the true distances in 3D, which is why one should rely on the lines that visualize the clusters (clustering was performed in 3D). Participant 7 and 8 are excluded, because no individual model could be fitted. In case of two fitting models per participant (1,2,4,6,10) the first model is denoted as a, the second as b.

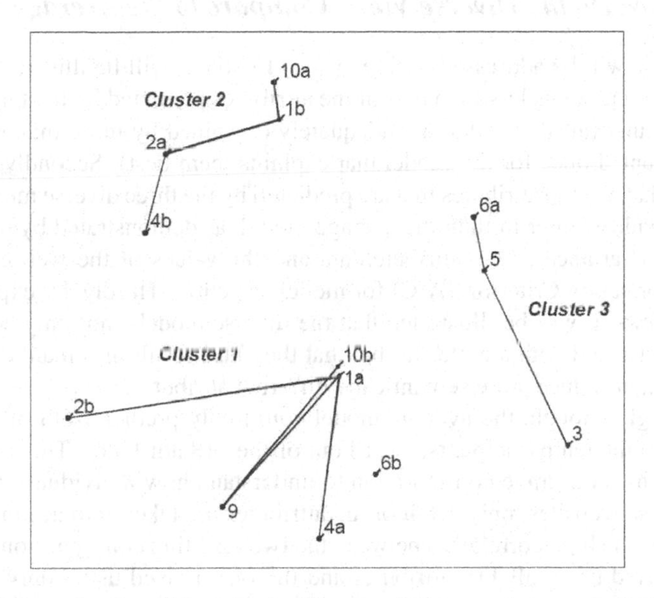

Fig. 3.3 A 2-D perspective of the 3-D visualization of distances between individual's different views.

Three clusters of models emerged. Cluster 1 summarizing 6 of the 13 single models (1a, 2b, 4a, 6b, 9, 10b), cluster 2 summarizing 4 models (1b, 2a, 4b, 10a) and cluster 3 summarizing the remaining 3 models (6a, 5, 3). The complementary models (a & b) for these five participants appear to be quite dissimilar as illustrated in figure 3.3 by the fact that they belong to different clusters. These clusters represent homogenous views, which can subsequently be mapped out.

3.3.5 Grouping the Homogeneous Views

In the last phase we establish a final set of configurations that represent the major diverse views across all participants and all attributes, on the set of stimuli. Views that belong in the same cluster are analyzed together and a shared MDS configuration is sought. Attributes that are not adequately predicted by the model are eliminated with the same criteria as in phase 1. The resulting 'averaged' views are then used for modeling the attributes from all participants. Attributes are allowed to exist in more than one configuration if they are adequately explained by all of them. When attributes in the same semantic category are not significantly different (which can be deduced from the fact that they have overlapping confidence ellipses in the K-dimensional configuration space), they are grouped. Attributes that cannot be grouped (have no replicates) are eliminated since no evidence exists that they contain reliable information.

3.3.6 *How Do the Diverse Views Compare to the Average View?*

This question will be addressed in three ways. Firstly, it will be illustrated that the average model predicts less than half of the attributes predicted by the three diverse models together (attributes that are adequately explained by more than one model are only counted once for the model that explains them best). Secondly, it will be illustrated that, for the attributes that are predicted by the three diverse models, these models provide a better fit than the average model, as demonstrated by the amount of explained variance in the attribute data and the values of the well established Akaike Information Criterion (AIC) for model selection. Thirdly, by exploring the resulting views, it will be illustrated that the diverse models, not only account for more attributes and with a better fit, but that they also result in semantically richer insights, i.e., introduce more semantically different attributes.

Surprisingly enough, the average model could only predict 1/6th of all the attributes from the ten participants, i.e. 18 out of the 118 attributes. This means, that when deriving an average configuration to understand how individuals distinguish between these websites, only 1/6th of the attributes are taken into account. This is illustrated by the high correlation between the two resulting configurations (R=*.99*), the one derived using all 118 attributes and the one derived using only the 18 attributes that are well predicted. Thus, the consequence of averaging is that we account only for 1/6th of the information available. The three diverse models predict 12, 10, and 16 attributes respectively (attributes predicted by more than one model were excluded from the ones that displayed the least fit). Thus, by accounting for diversity, even with our clearly heuristic and therefore sub-optimal procedure, we account for more than double the number of attributes than in the case of the average model.

Table 3.5 illustrates the goodness of fit of the average and the three diverse models for the 38 in total attributes resulting from models 1 to 3. As expected, a significant increase in the accounted variance (R^2) of the attribute data is observed as we the move from the average to the specific (i.e. diverse) model. But, does this increase in the goodness of fit of the model outweigh the increase in model complexity, i.e. going from one 2D to three 2D models? One of the most widely used criteria for model selection is the Akaike Information Criterion (AIC) (Burnham and Anderson, 2004) which is a function of the log likelihood value reflecting the goodness of fit of the model and the M degrees of freedom in the model reflecting its complexity:

$$AIC_c = -2\log(L(\hat{\theta})) + 2M\frac{n}{n-M-1} \tag{3.4}$$

Burnham and Anderson (2004) proposed a set of heuristics when comparing the AIC values of two models. Δ_i reflects the difference between the AIC of the

Table 3.5 Goodness of fit of the average and the three diverse models for the 38 in total attributes resulting from models 1 to 3

Attribute		R2			
No	*Variance* (R^2)	Average	Model 1	Model 2	Model 3
1	2	0.36	0.99		
2	4.3	0.86	0.99		
3	1.7	0.91	0.95		
4	4.6	0.91	0.94		
5	1.7	0.32	0.91		
6	2.7	0.76	0.9		
7	4.2	0.91	0.9		
8	6.6	0.7	0.87		
9	1.7	0.21	0.87		
10	6.1	0.83	0.86		
11	4.2	0.75	0.85		
12	3	0.68	0.69		
1	2.2	0.39		1	
2	5.6	0.78		1	
3	1.9	0.7		0.98	
4	6.3	0.82		0.94	
5	4.7	0.68		0.9	
6	2.8	0.54		0.9	
7	1.8	0.89		0.9	
8	3.8	0.68		0.89	
9	2.3	0.76		0.89	
10	1.9	0.54		0.88	
1	4.8	0.75			1
2	6	1			1
3	7	0.67			0.95
4	1.7	0.91			0.94
5	7	0.93			0.94
6	2.6	0.99			0.93
7	4.3	0.89			0.93
8	1.6	0.83			0.92
9	3.7	0.91			0.91
10	5.9	0.91			0.91
11	2.6	0.88			0.9
12	7.4	0.8			0.9
13	6.1	0.83			0.9
14	1.7	0.8			0.87
15	4.7	0.59			0.85
16	2.7	0.57			0.83
AIC_{avg} = 1480 AIC123 = 1126 Delta = 354					

simpler (i.e., average) model and the AIC of the more complicated one (i.e., consisting of three sub-models). $\Delta_i \leq 2$ provides significant evidence for the simpler model, $4 \leq \Delta_i \leq 7$ provides weak support for the heterogeneous model, while $\Delta_i \geq 10$ provides strong support for the heterogeneous model. In our case $\Delta_i = 354 \gg 10$, providing significant evidence that the diverse models, despite the increase in the model complexity, perform better than the average model.

Figure 3.4 illustrates the insights gained by the average and the three diverse models that are derived from the views corresponding to clusters 1 through 3. A significant overlap exists between models 1 and 3 (five common attributes), while model 2 provides a completely different view.

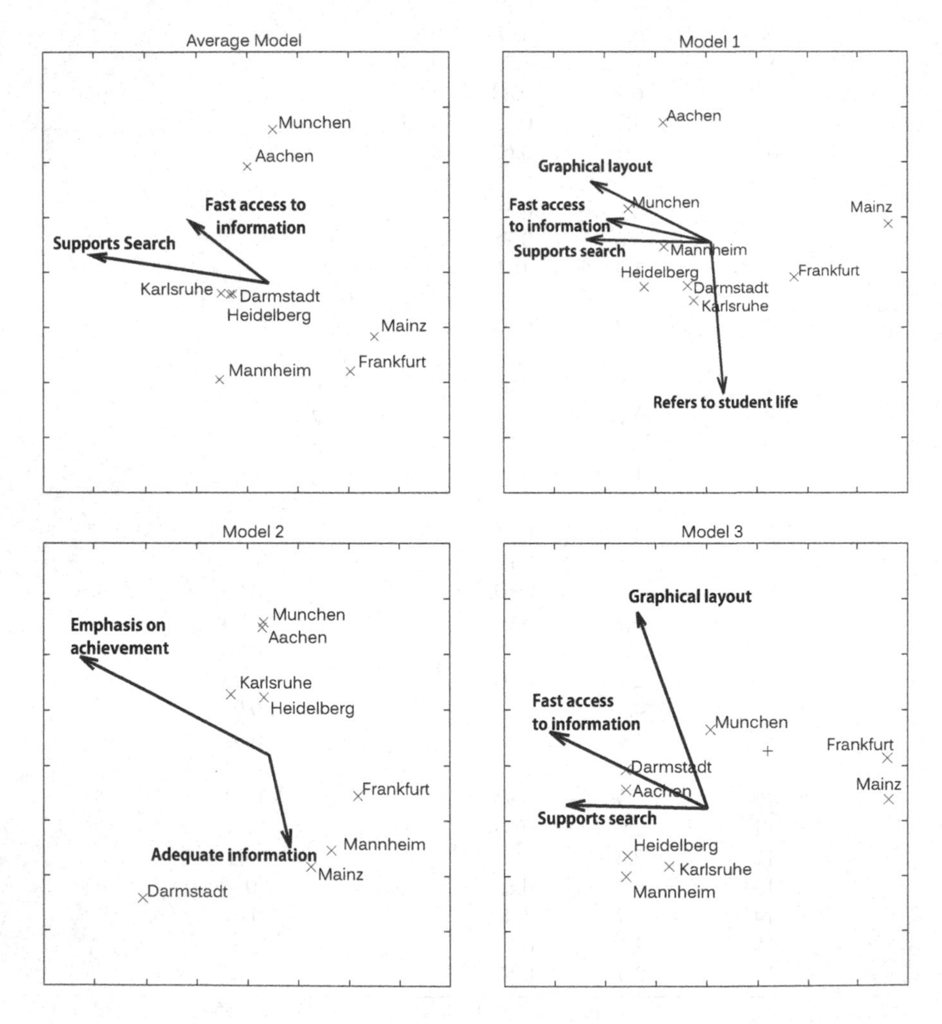

Fig. 3.4 The average model and the three diverse models corresponding to clusters 1 to 3 in figure 3.3.

The average model, although it accounts for more attributes than each of the diverse models, fails to predict semantically similar attributes. Thus, replicate attributes (i.e. attributes pointing towards the same direction with overlapping confidence ellipses) exist only for two attribute categories, namely *"Fast access to information"* and *"Supports search"*. The websites of the university of Mnchen and Aachen are differentiated from the remaining ones as websites that provide fast access to information, while the second attribute differentiates mainly the websites of the universities of Mainz and Frankfurt as the ones that do not support searching.

These two attributes are present also in two of the three diverse models, model 1 and model 3. Model 1 further differentiates the websites of Aachen and Mnchen as having a *"graphical layout"*, the website of the university of Aachen is mainly differentiated from all others as a website that does not *"refer to student life"*. On the contrary, model 2 provides a different insight. It reveals that the websites of the Universities of Mannheim, Frankfurt and Mainz put *"less emphasis on achievement"*. The set of websites can also be split in two groups based on the amount of information that they provide to the user.

3.4 Discussion

This chapter challenged traditional "averaging" practices in the analysis of repertory grid data. It was illustrated that only 15% of the attributes were adequately predicted by the average model and that the remaining attributes had minimal contribution to the configuration space, as shown by the high correlation (r=0.99) of the two spaces, i.e. one employing all attributes and one employing only the 15% of attributes being adequately predicted. The diverse models accounted for more than double the information accounted by the average model (38 attributes out of all 118), however a substantial amount of attributed remained non-modeled. One might wonder whether the information contained in the remaining attributes is truly non-modeled, or whether these attributes just do not differentiate strongly between the stimuli and thus do not contain any substantial information.

Figure 3.5a displays the variance in the attribute's ratings versus the accounted variance by the three diverse models for two groups of attributes: the ones that are adequately predicted by the three diverse models (38) and the remaining ones. One may note that the hypothesis stated above does not necessarily hold; no substantial differences may be found in the variance of the modeled and the non-modeled attributes. It thus becomes evident that while the diverse models result in a substantial increase in the information being modeled, still a substantial amount of attributes containing high variance remain non-modeled.

One potential reason for this might be a limitation that we pointed earlier, the use of heuristics in judging the goodness of fit of a model for a given attribute's data. An alternative procedure proposed by (Martens, 2009a) employs an iterative process, grounded on Singular Value Decomposition (Eckart and Young, 1936), that aims at optimizing the overall accounted variance (R^2) for all attributes in the data.

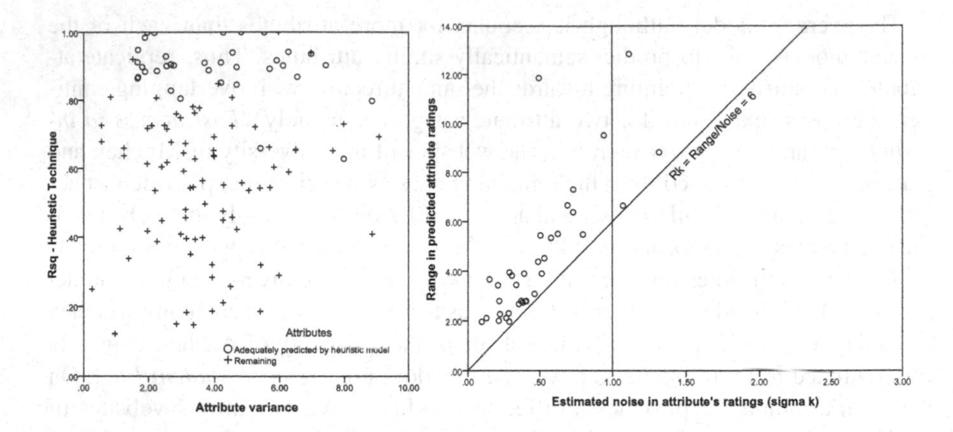

Fig. 3.5 (left) Scatter plot of the accounted variance by the three diverse models of the heuristic technique versus the variance in the original attribute ratings, and (right) scatter plot of the range versus the estimated noise in predicted attribute ratings.

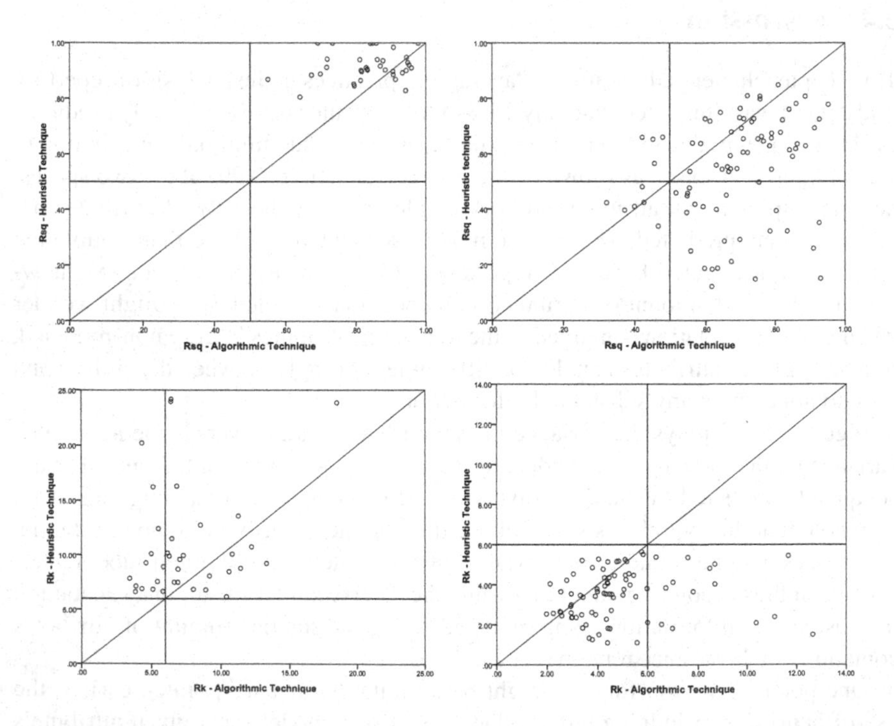

Fig. 3.6 R^2 (top) and R_k (bottom) values for the 38 attributes that were adequately modeled by the three models of the heuristic technique proposed in this chapter (left) and the remaining attributes (right), for the heuristic technique and the algorithmic technique proposed by Martens (2009a).

Figure 3.6 illustrates a comparison of the performance of the heuristic technique versus the algorithmic technique which explicitly optimizes the accounted variance R^2. As expected, the algorithmic techniques provides an improvement in terms of the accounted variance R^2 - in total 109 out of the 118 (92%) meet the explicit criterium R^2>.5 as compared to the 87 attributes (74%) of the heuristic technique. Nevertheless, the technique provides no improvement in terms of R_k (38 attributes meet R_k>6), which is somewhat expected as it does not explicitly take this criterion into account in the optimization process. Thus, an ideal technique, should take into account both R^2 and R_k in optimizing the goodness of fit of a model for a given set of attributes. While R^2 reflects the amount of variance in the attribute's ratings accounted for by the model, some attributes might contain only limited variance, and thus display high R^2 while providing limited meaningful differentiation between the stimuli. While our heuristic technique explicitly took into account the R_k, it has a number of limitations that could be addressed in the future. First, while R_k was used as a criterion in the determining whether an attribute is adequately predicted by a given model, it was not employed in the Multi-Dimensional Scaling optimization procedure for determining the best configuration of stimuli when trying to model a set of attributes. Thus, the majority of the attributes resulted in low R_k values. Second, as R_k is the range in the predicted attribute ratings divided by the estimated noise in the attribute's ratings, the denominator, i.e. estimated noise, will obviously have a significant impact on the R_k, and thus may render it as unreliable if we cannot reliably estimate the noise. In other words, if noise is unrealistically low, this will result in unrealistically high values of R_k. This issue pertains to the nature of the dataset as they lack repeated measures, thus noise cannot be directly estimated within a single participant and for a given attribute. Eliciting repeated measures in a repertory grid study would add substantial effort which might render the technique infeasible in several contexts. Figure 3.5b depicts the range $\hat{A}_{k,max} - \hat{A}_{k,min}$ versus the estimated noise (σ_k) for the 38 attributes that are adequately modeled by the heuristic technique. One may note that for the majority of the attributes σ_k<1, which is rather unrealistic given the quantization error when transforming the discrete data into continuous. Thus, an ideal technique should not allow for σ_k<1:

$$R_k = \frac{\hat{A}_{k,max} - \hat{A}_{k,min}}{min(1, \sigma_k)} \tag{3.5}$$

Given the limitations of the proposed R_k measure one might then question the validity of our initial result, that only 18 out of all 118 attributes are adequately modeled by the averaging analysis, since R_k was critical for determining goodness of fit. We earlier noted however a high correlation (r=.99) of the two configuration spaces, one attempting to model all 118 attributes and one attempting to model only the 18 best predicted attributes. This high correlation means that when excluding the 100 least-fit attributes, there was virtually no change in the MDS configuration space. In other words, when including these 100 least-fit attributes in the MDS optimization procedure, they did not contribute in the creation of this space. We also saw earlier that these attributes were not substantially different from the 18 best-fit in terms of the variance in their ratings. Thus, the averaging analysis results in a configuration space that is optimized for a small fraction of the dataset. We thus argue that despite the limitations of the R_k measure, averaging analysis fails to model a substantial fraction of the dataset.

3.5 Conclusion

This chapter argued against averaging in the analysis of personal attribute judgments. It was illustrated that when using averaging only 1/6th of the attributes in our study, i.e. 18 out of 118, were taken into account. A new MDS procedure that can better account for diversity in judgments was developed and its added value was illustrated through the reanalysis of published data. The analysis resulted in three diverse views on the data which were directly compared to the average view that is the common practice in RGT studies. The diverse models were found a) to account for more than double of the attributes accounted for by the average model, b) to provide a better model fit even for the attributes that were adequately predicted by the average model, and c) to result in semantically richer insights, since the diverse models can account for more semantically different attributes.

It was further illustrated that diversity exists not only across different individuals, but also within a single individual, in the sense that different attribute judgments of a participant may reveal different, complementary, views. At any point in time individuals can have different, seemingly conflicting views. For instance, individuals may regard one car as beautiful, but at the same time expensive. Individuals' overall evaluations of the car might thus be modulated by contextual aspects such as their motivational orientation (whether they just saw it in a newspaper on a Sunday morning or they are in the process of purchasing it, see Hassenzahl et al., 2008). Thus, being able to understand individuals' conflicting views is crucial for understanding how individuals infer the overall value of a product.

These insights strongly advocate the view that the analysis of quality judgments of interactive products should not stop on a group level, but must be extended to the relations between the attribute judgments within an individual. The Repertory Grid combined with the suggested technique to analyze the resulting quantitative data is an important step towards the adequate account of homogeneity and especially diversity in individual quality judgments.

Chapter 4
User Experience Over Time

Abstract. Product evaluation practices have traditionally been focusing on early interactions. Recent studies and trends in the consumer electronics industry, however, highlight the importance of understanding prolonged use. This chapter presents two studies that inquired into how users' experiences with interactive products develop over time. The first study assessed the ways in which ten individuals formed overall evaluative judgments of a novel interactive product at two moments in time during the adoption of a product, more precisely, in the first week and after four weeks. The second study followed six individuals through an actual purchase of a novel product and inquired into how their expectations and experiences developed from 1 week before until 4 weeks after the purchase of the product.

The chapter attempts two contributions. Firstly, it provides some initial insights into the differences between initial and prolonged experiences in terms of the way users form overall evaluative judgments about a product. Secondly, it raises a number of methodological issues in the assessment of the dynamics of user experience over time. This chapter ends with a proposal for a new approach to the study of the dynamics of experience over time and raises two research questions that will be addressed in chapters 5 and 6 respectively.

4.1 Introduction

Product evaluation practices have traditionally been focusing on early interactions. As a result, we have been mostly concerned about product qualities relating to initial use. Den Ouden et al. (2006), however, highlighted that the reasons for product returns span a wider range of aspects than just problems relating to the learnability and usability of interactive products. Moreover, a number of recent trends are highlighting the importance of longitudinal evaluation practices. First, legislation and competition within the consumer electronics industry has resulted in an increase in the time-span of product warranties. This has resulted in an alarmingly increasing number of products being returned on the basis of failing to satisfy users' true needs

E. Karapanos: Modeling Users' Experiences with Interact. Syst., SCI 436, pp. 57–83.
springerlink.com

(Den Ouden et al., 2006). Secondly, products are increasingly becoming service-centered. Often, products are being sold for lower prices and revenues are mainly coming from the supported service (Karapanos et al., 2009c). Thus, the overall acceptance of a product shifts from the initial purchase to establishing prolonged use.

This chapter attempts two contributions. First, it attempts to inquire into the differences between initial and prolonged experiences in terms of the way users form overall evaluative judgments about products across time. In the field of user experience, a number of studied have attempted to inquire into how users form overall evaluative judgments of products on the basis of quality perceptions (Hassenzahl, 2004; Mahlke, 2006; Tractinsky and Zmiri, 2006; Hartmann et al., 2008; Van Schaik and Ling, 2008). An aspect that has been largely overlooked so far is that of temporality, i.e. how users' experiences develop over time. As we use a product, our perception of the qualities of the product will change (von Wilamowitz Moellendorff et al., 2006). For example, we get used to it, which eventually changes our perception of its usability; at the same time it excites us much less than in our first moments with it. Even more interestingly, at different phases of use we will evidently attach different weights to different qualities. In our first interactions with a product we may focus on its usability and the stimulation that it provides to us. After we use it for some time, we might become less concerned about its usability, and other aspects of the product such as novel functionality or communication of a favorable identity to others become more important. Two studies are being reported. In study 1 we provided 10 participants with an innovative pointing device connected to an Interactive TV set top box, and elicited their perceptions during the first week as well as after four weeks of use. In study 2 we followed six participants through the purchase of an Apple iPhone and the first month of use. Both studies provide empirical findings on the differences between initial and prolonged experiences. A conceptual model of temporality of experience is attempted and implications for HCI practice are suggested.

Second, it raises a number of methodological issues in the assessment of the dynamics of user experience over time. It questions the value of reductionist approaches where a-priori defined measurement models are employed in measuring the user experience and suggests an alternative methodological approach that relies on a) eliciting the experiences that are personally meaningful to each participant, in the form of *experience narratives*, and b) employing content analysis techniques in creating multiple levels of abstraction, from concrete idiosyncratic insights to abstracted and generalized knowledge.

4.2 Background on Experience and Temporality

This section reviews some holistic and reductionist models in user experience research and discusses their relevance for the study of the temporality of users' experiences.

4.2.1 Temporal Aspects in Frameworks of Experience

The holistic thread of user experience has contributed a number of frameworks describing how experience is formed, adapted, and communicated in social contexts. Forlizzi and Battarbee (2004) described how experience transcends from unconsciousness to a cognitive state and finally becomes "an experience", something memorable that can also be communicated in social interactions. Battarbee and Koskinen (2005) elaborated on the social mechanisms that lift or downgrade experiences as they participate in our social interactions. McCarthy and Wright (2004) described how sense-making takes place in the development of experience by decomposing it into six processes, from anticipation to reflection and recounting.

Although one can note that these frameworks approach temporality through a micro-perspective, i.e. how experiences are formed, modified and stored, one could also raise a number of macro-temporal issues. For instance, does the distribution between unconscious and cognitive experiences remain stable over time or do cognitive experiences reduce as users' familiarity increases? Next, what motivates the process of lifting up experiences and communicating them in social contexts? Do these underlying motivations change over time, e.g. as users' initial excitement fades out? A framework of temporality of experience, proposed in this chapter, aims at providing answers to these questions by conceptualizing the missing dimension of time.

4.2.2 Beauty, Goodness and Time

Reductionist approaches to user experience have contribute a wealth of new measurement and structural models. Hassenzahl (2004) distinguished between two quality perceptions: *pragmatic* and *hedonic*. Pragmatic quality, he argued, refers to the product's ability to support the achievement of behavioral goals (i.e. usefulness and ease-of-use). On the contrary, hedonic quality refers to the users' self; it relates to *stimulation*, i.e. the product's ability to stimulate and enable personal growth, and *identification*, i.e. the product's ability to address the need of expressing one's self through objects one owns. Tractinsky and Zmiri (2006) drew on the work of Rafaeli and Vilnai-Yavetz (2004) to propose three distinct product quality attributes: *usability*, *aesthetics* and *symbolism*. Forlizzi (2007) extended this model to further account for the emotional and social aspects of product use.

An interesting question relates to how these quality perceptions are combined to form an overall evaluation of the product (Hassenzahl, 2004; Mahlke, 2006; Tractinsky and Zmiri, 2006; Hartmann et al., 2008; Van Schaik and Ling, 2008). Hassenzahl (2004) suggested two distinct overall evaluative judgments of the quality of interactive products: *beauty* and *goodness*. He found goodness to be affected primarily by pragmatic aspects (i.e. usefulness and usability). On the contrary he found beauty to be a rather social aspect, largely affected by identification (i.e. the product's ability to address the need of self-expression). In a similar vein, Tractinsky and Zmiri (2006) distinguished between *satisfying* and *pleasant* experiences. They found perceptions of usability to be better predictors for a satisfying rather

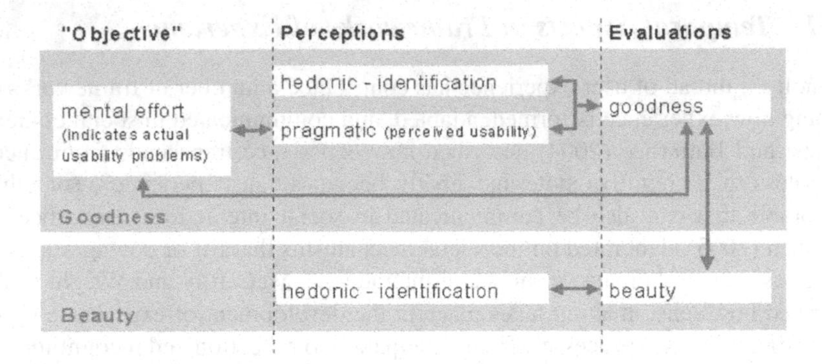

Fig. 4.1 Hassenzahl's (2004) framework distinguishing between two overall evaluative judgments, i.e. goodness and beauty, and three quality perceptions, i.e. pragmatic quality, hedonic quality - stimulation, and hedonic quality - stimulation.

than pleasant experience while perceptions of the products' aesthetics to be better predictors for a pleasant rather than satisfying experience.

But, how stable are such relations over time? Study 1 attempts an initial insight in the differences between initial and prolonged experiences in the way users form overall judgments about products. In Hassenzahl's study the users' experience with the products was limited to short interaction episodes, where users were asked to carry out a number of predefined tasks. Study 1 attempts to explore how the user experience and the subsequent evaluative judgments develop over a longer period of time, and in less controlled interactions.

4.3 Study 1

The objective of study 1 is to understand how users form evaluative judgments during the first experiences with a product and after prolonged use. Hassenzahl's (2004) distinction between the two evaluative judgments of *goodness* and *beauty* was employed in an effort to replicate existing findings and extend them over prolonged use. Given previous results one would expect beauty to be relatively stable over time (Lindgaard et al., 2006; Tractinsky et al., 2006; Hassenzahl, 2004) and related to the self-image (Hassenzahl, 2004; Tractinsky and Zmiri, 2006; Tractinsky et al., 2006) that the product communicates to relevant others. Further, one would expect judgments of goodness to be primarily affected by the product's pragmatic aspects (i.e. utility and usability) (Hassenzahl, 2004; Mahlke, 2006).

4.3.1 Method

4.3.1.1 Participants

A total of ten individuals (four female) participated in a four weeks study of an Interactive TV set-top box (STB). They all responded to an invitation that was placed on the website of a multinational consumer electronics company. Their age varied from 22 to 35 years (mean 26y). Their likelihood to recommend the brand ranged from 3 to 9 on a 10-point scale (mean 7.8, std. 2.3). Participants were classified to respective market segments based on demographic information. A bias towards innovator consumers was observed, as it was expected. The study focused on a particular part of the set-top box, uWand. uWand is a novel pointing device for interactive TV contexts. It uses led technology to identify where the user points at within the content that appears on the TV.

Fig. 4.2 A novel pointing device for Interactive TV context, used in Study 1

4.3.1.2 Procedure

During the four week testing period participants were asked to rate uWand at two different times, during the first week of use as well as at the end of the 4th week. The AttracDiff 2 questionnaire (Hassenzahl, 2004) was employed for the assessment of three distinct aspects of the quality of interactive products: *pragmatics* (e.g. simple, practical, clear), *stimulation* (e.g. innovative, exciting, new) and *identification* (e.g. inclusive, classy, presentable). Each quality aspect is measured with seven bipolar attributes, employed in Semantic Differential scales (Osgood et al., 1957). Beauty and goodness were measured with single items (taken from AttracDiffs appeal construct). Both are evaluative judgments with Goodness focusing on the complete product, while Beauty is rather restricted to visual features. Note that for evaluative, high level summary judgments single item measurements are appropriate and commonly used (e.g., to measure subjective well-being).

4.3.2 Results

Since we were interested in a detailed picture of the relationship between product attributes, we decided to analyze every attribute (in Table 4.1) separately. The distance

$D_{ij} = 1 - —R_{ij}—$ between the individual attributes i and j was derived from the correlation R_{ij} between the ratings by different participants on the two attributes. The obtained distances were subsequently visualized in three dimensions (Stress value S=0.19 1st week; S=0.15 4thweek) using the Multidimensional Scaling tool XGms (Martens, 2003). Hierarchical clustering (with minimum variance) was applied to the predicted distances in the 3-dimensional space. Figure 4.3 illustrates a 2D projection of the 3-dimensional visualization of the distances between the quality attributes. The obtained clusters are denoted by the connecting lines. The left figure reflects the users' ratings during the first week of use while the right figure reflects the users' ratings after four weeks of use. All in all, clusters derived from the measurement in the first week reflect the assumed underlying constructs, with a close knit groups of mainly pragmatic and hedonic stimulation attributes and a looser rest of hedonic identification attributes. After four weeks, the perceptions seem much more differentiated, and relationships among attributes lost some of their strength.

Table 4.1 Bipolar attributes measuring *pragmatic quality*, *stimulation* and *identification* (Hassenzahl, 2004)

PragmaticQuality	***Stimulation***	***Identification***
Technical-Human	Typical-Original	Isolating-Integrating
Complicated-Simple	Standard-Creative	Amateurish-Professional
Impractical-Practical	Cautious-Courageous	Gaudy-Classy
Cumbersome-Direct	Conservative-Innovative	Cheap-Valuable
Unpredictable-Predictable	Lame-Exciting	Noninclusive-Inclusive
Confusing-Clear	Easy-Challenging	Takes me distant from people - Brings me closer to people
Unruly-Manageable	Commonplace-New	Unpresentable-Presentable

During the first week, Beauty judgments relate mostly to attributes reflecting the quality of stimulation (i.e. original, creative, new, innovative) and to one attribute reflecting identification (i.e. classy). This close relationship between stimulation and beauty seems to have disappeared after four weeks of use; beauty now seems to be a disjoint judgment only related to one identification (i.e. classy) and one stimulation (i.e. challenging) attribute. While the relation between "classy" and beauty remained over the period of four weeks, stimulation seemed to influence beauty judgments mostly during the first experiences. Surprisingly, stimulation seemed to be a more important determinant of beauty than identification in the first experiences.

A similar pattern can be observed for judgments of Goodness. During the first week judgments of goodness relate mostly to pragmatic quality attributes (i.e. practical, direct, manageable, predictable, clear) and to one attribute reflecting identification (i.e. presentable). After four weeks of use goodness appears to be related mostly to identification (i.e. professional, inclusive, valuable, integrating, brings me

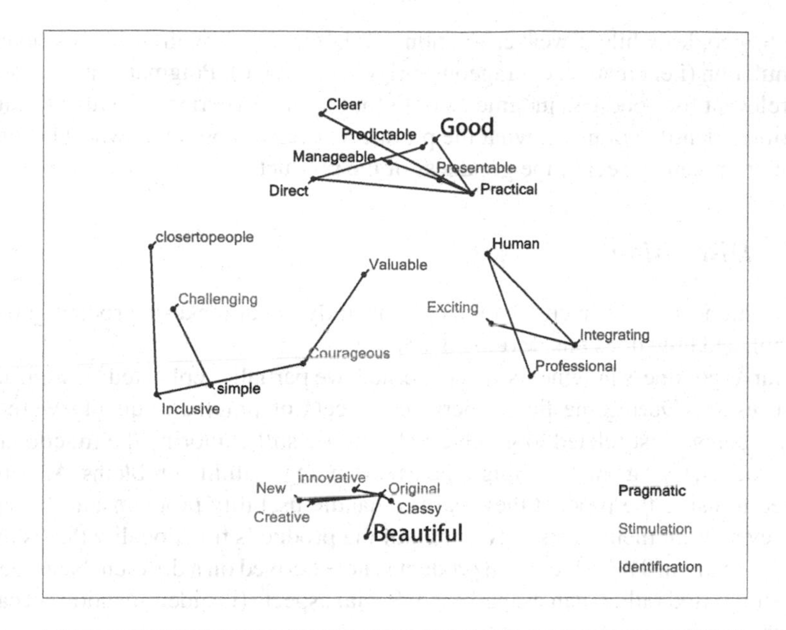

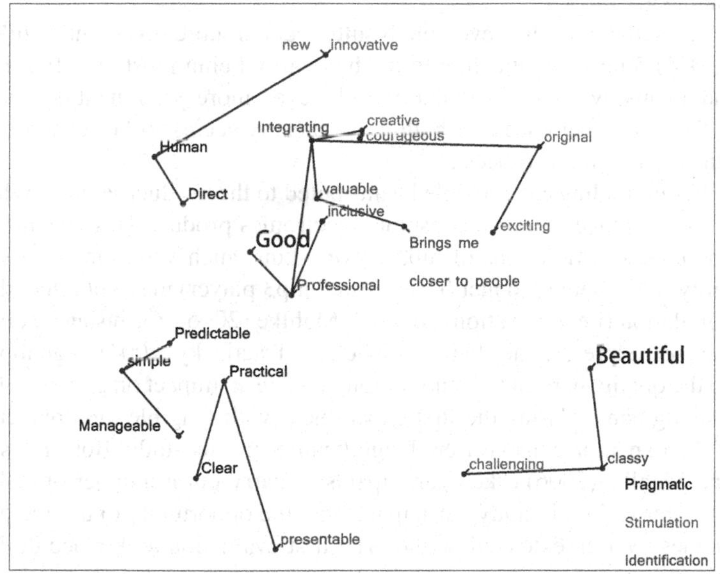

Fig. 4.3 2D view on the 3-dimensional visualization of distances between quality attributes, beauty and goodness. Users' perceptions during the 1st week of use (left) and after 4 weeks of use (right).

closer to people) while a weaker relation can be observed with attributes relating to stimulation (i.e. creative, courageous, original, creative). Pragmatic aspects seem to be relevant for goodness judgments only for the first experiences with a product. Over time, identification (i.e. what the product expresses about its owner) becomes a more prominent aspect of the goodness of the product.

4.3.3 Discussion

Two questions were of interest in the current study: what makes a product good or beautiful, and how does this develop over time.

As far as goodness judgments are concerned, we partially replicated Hassenzahl's (2004) results. During the first experiences facets of pragmatic quality were the qualities being most related to goodness. Users are still exploring the functionality of the product, trying out new things and experiencing usability problems. As people get used to using the product they learn to handle usability problems; at the same time they restrain themselves only to part of the product's functionality that is most attractive to them. The value of the product is now derived on a different basis, being ownership-based rather than usage-based. Social aspects (i.e. identification) became more prominent.

For beauty judgments, however, the results seemed more divergent. While Hassenzahl (2004) found identification to be the quality being most closely related to beauty judgments, we found stimulation to be even more prominent than identification in the first experiences. In both cases, beauty seems to be related more to hedonic than to pragmatic aspects.

This different finding can possible be attributed to the product sample. While the current study employed a novel consumer electronics product, Hassenzahl's study focused on different variations of mp3 player skins; such variations in aesthetics and usability of the same product (i.e. a virtual mp3 player) may not affect the perceived stimulation (i.e. innovation, novelty). Mahlke (2006), for instance, observed a quality called expressive aesthetics (Lavie and Tractinsky, 2004), arguably comparable to the quality here called stimulation, to have an impact on goodness but not on beauty judgments, during the first experiences with tangible mp3 players. The nature of the experience also differed significantly in this study. Both Hassenzahl (2004) and Mahlke (2006) asked participants to carry out a number of tasks in a laboratory context; in this study participants had the opportunity to use the product at their homes over an extended period. The first evaluation took place during the first days of use.

After four weeks of use, stimulation seemed to loose its dominance on beauty judgments. Eventually, users were not any more surprised by the product's stimulating character and the product's novelty lost its power to make the product more beautiful in the users' eyes.

Overall, despite the exploratory character of the study, it seems that we came across some interesting results. The proposition that novelty and stimulation impact beauty judgments resonates with Berlyne's work on stimulation and surprise as arousal-heightening attributes and their impact on the appraisal of art (Machotka, 1980). Furthermore, time seems to have an impact on the importance we attach to different qualities of the experience with interactive products. For example, despite the crucial importance of usability in a product's initial acceptance, aspects of product ownership (and not use) are essential for a user to resonate with a product and value it in the long term.

4.3.4 *Limitations of the Study*

A number of limitations of the study have to be noted before proceeding to general conclusions. First, an inherent limitation of the study was the evaluation of only one product and only by a small number (10) of participants. As a result, these findings cannot be generalized to a wider population and range of products; they provide, however, some initial insights that can be falsified or corroborated in subsequent studies with different products and participants. However, it highlights an inherent limitation of longitudinal studies in producing generalized knowledge due to their labor intensive nature which evidently restricts the sample of products, participants, or time studied. Alternative techniques such as the one we will propose in the next chapter, and others proposed by von Wilamowitz Moellendorff et al. (2006) and Fenko et al. (2009) are essential for overcoming the labor-intensive nature of longitudinal studies.

Second, the two measurements (1_{st} and 4_{th} week of use) assessed users' perceptions at each time (current state) rather than directly assessing how their perceptions changed over time. One could be concerned about the sensitivity as well as the reliability of such absolute measures where judgments do not take place in contrast to some specified reference point. As discussed in chapter 2, paired comparison scales that define explicitly both reference points, have been shown to be less sensitive to contextual effects than single-stimulus scales. Thus, one would assume that by explictly asking participants to assess how their opinion on a given quality has changed over the course of time, may increase the technique's sensitivity in assessing this change.

Third, one major limitation of the study was the use of a-priori defined measurement model in a pre-post design (with only two measures being taken). Although in such a study design, measurement has the least influence on users' attitudes, it provides a rather limited insight into the actual patterns of change over time. First, as only two measures were taken, one could not know whether the identified changes were an effect of time, or of random contextual variation. Second, as we employed an a-priori defined measurement model (Hassenzahl, 2004), we might have evidently missed constructs that dominate prolonged use and are not apparent in this model. Secondly, due to the quantitative nature of the study, we gained no insight as to the reasons for these changes in users' experiences. In study 2, we attempt a qualitative understanding into the differences between initial and prolonged use.

4.4 Study 2

Study 1 provides some evidence for the point of view that prolonged experiences with products can differ substantially from initial experiences in terms of the way that different product qualities relate to each other.

The question that was raised then was: what causes these changes? Can we describe the adoption of a product in terms of distinct phases? And what qualities would dominate each of these phases? While longitudinal studies on product adoption are scarce in the field of HCI, much work has been performed in the field of cultural studies of technology (Du Gay et al., 1997; Silverstone and Haddon, 1996), trying to understand how technology is being adopted and incorporated in specific cultures. We agree with McCarthy and Wright (2004) that cultural studies have a tendency to downplay the role and diversity of individual experience, yet, we believe that much can be learned from examining the relevance of cultural studies frameworks for the study of user experience.

A promising framework for the study of prolonged user experiences is the one from Silverstone and Haddon (1996) on the dimensions of adoption. They suggested three dimensions, but also moments, in the process of technology adoption: *commodification*, *appropriation* and *conversion*. Commodification, they argued, refers to all activities from both producers and users that result in specific claims for a function and an identity for a new product. As users participate in the commodification process, they form expectations about ways in which the product could become relevant to their lives. In appropriation, users accept enough of the relevance of the product and they gradually incorporate it into their life routines. Finally, in conversion, users accept the product as part of their self-identity and employ it in their social interactions.

Silverstone and Haddon's framework, however, approach product adoption from a cultural and macro-temporal perspective, thus undermining the details that describe how individuals' experiences develop over time. For instance, commodification is conceived as an iterative process where both users and producers make claims for new functions, eventually resulting in new products in the market. They are less concerned about how expectations impact users' experience with a product. Next, how exactly does appropriation happen? As it will become evident later, we distinguish between two aspects of appropriation, namely orientation and incorporation.

This study, inspired by the framework of Silverstone and Haddon (1996), uses the iPhone to validate distinct phases in users' experience, and understand what differentiates them, how users' experience changes across these phases, and how this impacts users' evaluative judgments about the product. More specifically, it addresses the following questions:

1. Can users' experiences be articulated in distinct phases in the adoption of the product?
2. What motivates the transition across these phases?
3. How does each phase contribute to the overall perceived quality of the product?

4.4.1 *The Study*

4.4.1.1 Product

We selected the iPhone as a product of study because of its proven success not only during initial but also over prolonged use. This was considered important as we wanted to elicit experiences relating to the successful adoption of a product in the different phases as identified by Silverstone and Haddon (1996). The iPhone was moreover considered an interesting example as it highlights some non-instrumental aspects of experience (e.g. stimulation & identification (Hassenzahl, 2004)) that are currently discussed in the field of user experience.

4.4.1.2 Participants

We recruited 6 participants through a prescreening virtual advert of an iPhone sale. Our motivation was to recruit participants that were at that time considering the purchase of the product before motivating them to participate in the study with a monetary incentive. After responding to the advert, a second email was sent, introducing the details of the study and inviting them to participate. We observed a strong bias for participants with technical background. In the final selection we aimed for a homogeneous participant sample; only one participant did not previously own a smart phone. Their age ranged from 28 to 33 years (mean 31y). Two out of six were female.

4.4.1.3 Method

Our criteria for choosing a method were a) its ability to retrieve accurate recalls on the product's perceived quality within single experiential episodes, and b) its ability to elicit rich qualitative accounts on the experienced episodes. We chose the Day Reconstruction Method (DRM) (Kahneman et al., 2004; Schwarz et al., 2008) over the more popular Experience Sampling Method (ESM) (Hektner et al., 2007) and event-contingent diaries (Bolger et al., 2003), as it enables capturing rich qualitative accounts offline.

The DRM is typically conducted at the end of a reported day or at the beginning of the next day. In an effort to minimize retrospection biases, DRM asks participants to mentally reconstruct their daily experiences as a continuous series of episodes, writing a brief name for each one. Experiential episodes are thus being recalled in relation to preceding ones, which enables participants to draw on episodic memory when reporting on the felt experience (Schwarz et al., 2008). Hence, participants are better able to reflect on the perceived quality of the product within a single experiential episode, avoiding inferences from their global beliefs about the product. As demonstrated by Kahneman et al. (2004), the DRM combines the advantages of an offline method with the accuracy of introspective approaches such as the Experience Sampling.

4.4.1.4 Process

One week before the purchase of the product, participants were introduced to the study. During this week, participants were asked to capture their major expectations about the product in the form of short narratives. The perceived importance of each expectation was assessed, using a 7-point Likert scale, both before the purchase as well as at the end of the study. After purchase, participants captured their daily experiences at the end of each day. This process consisted of two main activities: *day reconstruction*, and *experience narration*. In day reconstruction, participants listed all activities of the day that somehow related to their iPhone. A brief name and an estimation of time spent were recorded for each activity. In experience narration, participants were asked to pick the *three most impactful*, either satisfying or dissatisfying, experiences of that day. They were explicitly instructed to *"use [their] own feeling or a definition of what 'satisfying' and 'dissatisfying' experience means"*. For each of the three experiences, participants were asked to write a story that describes in detail the situation, their feelings and their momentary perceptions of the product.

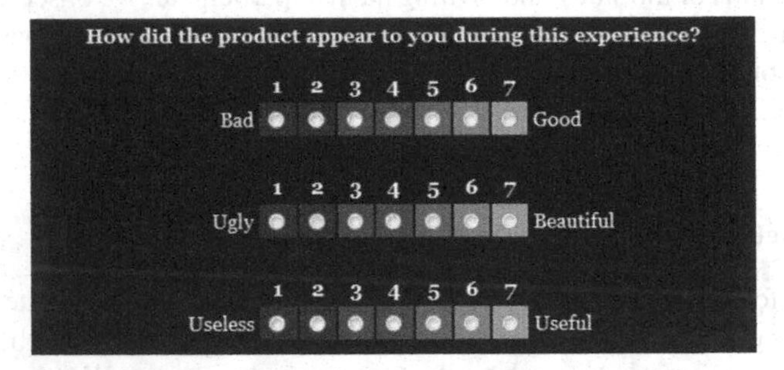

Fig. 4.4 Interface used for rating the product after reporting an experience

Finally, for each experience narration, participants rated the product as perceived within that specific situation (see figure 4.4. A shortened version of the Attrakdiff 2 (Hassenzahl, 2004) questionnaire was employed, that identifies two overall evaluative judgments, i.e. *beauty* and *goodness*, and three distinct product qualities: *pragmatics* (i.e. utility and ease-of-use), *stimulation* (i.e. the product's ability to address the human need of stimulation, novelty and challenge) and *identification* (i.e. the product's ability to address the need of expressing one's self through objects one owns). Each construct was measured with one single item (see table 4.2) that displayed the highest loading on the latent construct during a prior study (Karapanos et al., 2008a). Pragmatic quality was split in two distinct components, usefulness and ease-of-use. Identification was measured with a single item derived from Tractinsky and Zmiri (2006).

Table 4.2 Latent constructs measured through a single scale

Latent construct	Measurement item
Usefulness	useful - useless
Ease-of-use	simple - complicated
Stimulation	innovative - conservative
Identification	positive message about me - negative message about me
Goodness	good - bad
Beauty	beautiful - ugly

4.4.2 Data Analysis

A total of 482 experience narratives were collected during the four weeks of use. These were submitted to a conventional qualitative Content Analysis (CA) (Hsieh and Shannon, 2005; Krippendorff, 2004). Conventional CA is appropriate when prior theory exists but the researcher wishes to stay open to unexpected themes and only at a later stage relate findings to existing theory, whilst it shares a similar analytical approach with Grounded Theory. Our approach consisted of three steps:

Open coding - A detailed coding aimed at identifying key themes in the data without imposing pre-conceived categories. The process resulted in about 70 loosely connected codes referring to about 700 instances in the data.

Axial coding - In the second step, the initial set of phenomena described by open codes was categorized using axial coding. Open codes were grouped into categories which were subsequently analyzed in terms of properties and dimensions. This resulted in a set of 15 main categories reflecting aspects like the aesthetics of interaction, learnability and long-term usability (see table 4.3).

Quantitative analysis - All experience narratives were classified as being primarily related to one of the fifteen categories. This process was independently conducted by the first author and an additional researcher (Interrater agreement K=.88). Both researchers were already immersed in the data as they both participated in the axial coding process. Narratives for which no agreement was attained were excluded from the subsequent analysis. We avoided clarifying disagreements to ensure high uniformity within experience groups. The distribution of experience narratives over the four weeks of the study was then identified for each of the 15 categories. Based on the resulting temporal patterns and semantic information, the 15 categories were then mapped into 3 broad themes reflecting distinct phases in the adoption of the product: ***Orientation***, ***Incorporation*** and ***Identification***. An additional theme, called ***Anticipation***, was added to reflect users' a priori expectations that were captured during the first week of the study. Finally, separate regression analyses with the two overall evaluative judgments, i.e. *goodness* and *beauty*, as dependent and the four quality attributes, i.e. *usefulness*, *ease-of-use*, *stimulation* and *identification*, as independent variables, were run for the three main groups of experiences,

Table 4.3 The coding scheme of the analysis

Orientation	Incorporation	Identification
	Satisfying experiences	
1. Visual aesthetics	4. Long-term usability	9. Personalization
2. Aesthetics in interaction	5. Fast access to information	10. Daily rituals
3. Learnability	6. Alleviating boredom in idle periods	11. Self-expression
	7. Capturing momentary information	12. Sense of community
	8. Avoiding negative social situations	
	Disatisfying experiences	
13. Learnability problems	14. Long term usability problems	
	15. Usefulness problems	

i.e. Orientation, Incorporation, and Identification, to understand what product qualities dominate in each phase of use.

4.4.3 Findings

All in all, three phases were identified in the adoption of the product, i.e. *Orientation*, *Incorporation*, and *Identification*. These phases reflected different qualities of the product, which were found to display distinct temporal patterns. We conceptualized temporality of experience as consisting of three main forces, i.e. an increasing *familiarity*, *functional dependency* and *emotional attachment*. These forces motivate the transition across the three phases in the adoption of the product (figure 4.5).

Anticipation, i.e. the act of anticipating an experience resulting in the formation of expectations, happens prior to any actual experience of use. Micro-temporality, i.e. the emergence of a single experiential episode, is thus visualized as the transition from the core of the circle towards its outer radius. Our interactions are typically filled with a multitude of such experiential episodes. Each of these experiences highlights different qualities of the product such as its aesthetics or its daily usefulness. While many different experiences may co-exist in a single time unit (e.g. day), their distribution changes over time, reflecting the transition across different phases in the adoption of the product.

Orientation refers to users' initial experiences that are pervaded by a feeling of excitement as well as frustration as we experience novel features and encounter learnability flaws. In *Incorporation* we reflect on how the product becomes meaningful in our daily lives. Here, long-term usability becomes even more important than the initial learnability and the product's usefulness becomes the major factor

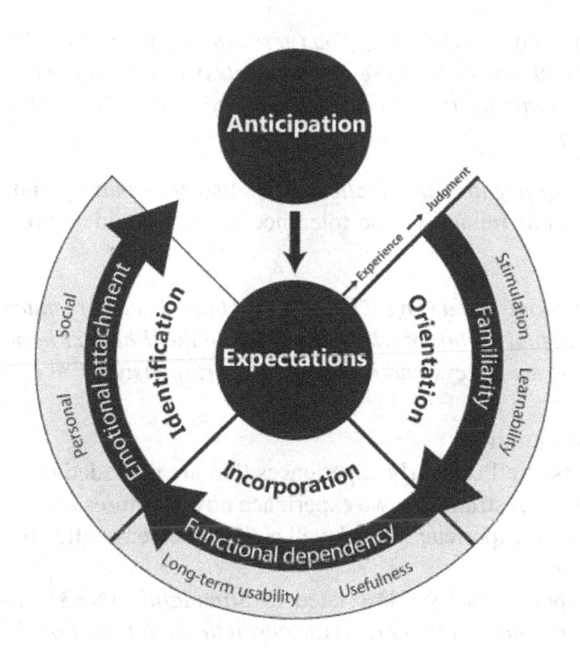

Fig. 4.5 Temporality of experience, consisting of three main forces, an increasing familiarity, functional dependency and emotional attachment, all responsible for shifting users' experience across three phases: orientation, incorporation and identification. In each phase, different product qualities are appreciated

impacting our overall evaluative judgments. Finally, as we accept the product in our lives, it participates in our social interactions, communicating parts of our self-identity that serve to either differentiate us from others or connect us to others by creating a sense of community. This phase we call *Identification*.

Next, we illustrate how this framework was developed from the actual study by addressing our three overall questions:

4.4.3.1 Can Users' Experiences be Articulated in Distinct Phases in the Adoption of the Product?

4.4.3.1.1 Anticipation

Participants formed an average of six pre-purchase expectations. Expectations related to *opportunities for positive experiences* (76%) such as the performance of the multi-touch screen, the incorporation of a mobile agenda and mobile internet in daily life, the aesthetics of packaging and product, as well as friends' and colleagues' reactions,

> "*...I bought my iPod not only as a music player but also as an organizer. But synchronizing iPod with my iCal was not that easy and I could not even add*

anything to my agenda using iPod (very bad of Apple). The iPhone will make my life much easier because of its seamless integration with Mac's iCal. I can add events using both devices and they will talk to each other as two natives talk..."

but also to *fears of negative implications* (24%) such as battery endurance, typing efficiency, as well as reliability and tolerance in daily accidents (e.g. drop on the ground):

"My last phone had a QWERTY keyboard that I liked very much. I am curious how the virtual keyboard will be working on the iPhone. I hope it's not going to have too small keys and it will be really responsive."

4.4.3.1.2 Orientation

Orientation refers to all our early experiences that are pervaded by a feeling of excitement as well as frustration as we experience novel features and encounter learnability flaws. These experiences displayed a sharp decrease after the first week of use (see figure 4.6).

Satisfying experiences (N=71) related to *Stimulation* (N=33) induced by the product's *visual aesthetics* (N=12) and the *aesthetics in interaction* (N=21), but also to positive surprises regarding the simplicity with which certain initial tasks could be carried out, i.e. *learnability* (N=38):

[Visual aesthetics, day 1] *"my first impression when I saw the box was WOW!, very nice!!", [Aesthetics in interaction, day] "when I clicked on the album, I just loved the way it turned around and showed all the songs in it",* ***[Learnability, day 2]****, "I tried to set up my iPhone's WiFi which I expected would be a little bit difficult... it was just 3 steps away! amazing! 3 steps away! It automatically detected the WLan and then connected to it. My iPhone was ready for internet browsing in less than a minute. Just cool!!!"*

Dissatisfying experiences reflected *learnability problems* (N=50) induced by unexpected product behavior:

" [day 3] "I started typing an SMS in Polish and the dictionary tried to help me by providing the closest English word. There was no button to switch the dictionary off, no easy option to edit my preferences about it."

4.4.3.1.3 Incorporation

As participants gradually incorporated the product in their lives, their experiences increasingly reflected the ways in which the product was becoming meaningful in diverse use contexts (see figure 4.6).

Satisfying experiences (N=113) related to design aspects that enhanced users' efficiency over time, i.e. *long-term usability* (N=43), but also to the product's *usefulness* (N=70), reflecting ways in which the product supported participants' daily activities. These related to providing *fast access to information* (N=33) when mobile, or at home, by *alleviating boredom in idle periods* (N=18) through activities such as browsing the web, browsing photos or playing games, by enabling capturing momentary information (N=11) when mobile, either probed by external stimuli or during introspection, and by *avoiding negative social situations* (N=8), e.g. when identifying typed phone numbers from contact list before establishing a call, enabling easy access to destination time when calling abroad, or allowing a fast mute of all sounds when being in a meeting:

> ***[Long-term usability, day 3]*** *"turning the iPhone sideways not only turns the page but also magnifies it, so text is easily readable. Truly well done! I don't see this kind of attention to details too often",* ***[fast access to information, day 3]*** *"it's so easy to just pick up the phone to check the web rather than having to switch the computer on - I am becoming a great fan of it. It's simply saving time",* ***[alleviating boredom in idle periods, day 7]*** *"I like playing - I find it a nice activity when waiting, traveling and at any point when I can't really do anything else",* ***[capturing momentary information, day 12]*** *"Now I tend to go joking when I want to think of my work as I can easily write down whatever comes to my head",* ***[avoiding negative social situations, day 22]*** *"It was so nice that iPhone recognized a phone number from my contacts list and showed it to me before I started calling. Thanks to that I didn't leave yet another voice message that would be staying there for another week or two."*

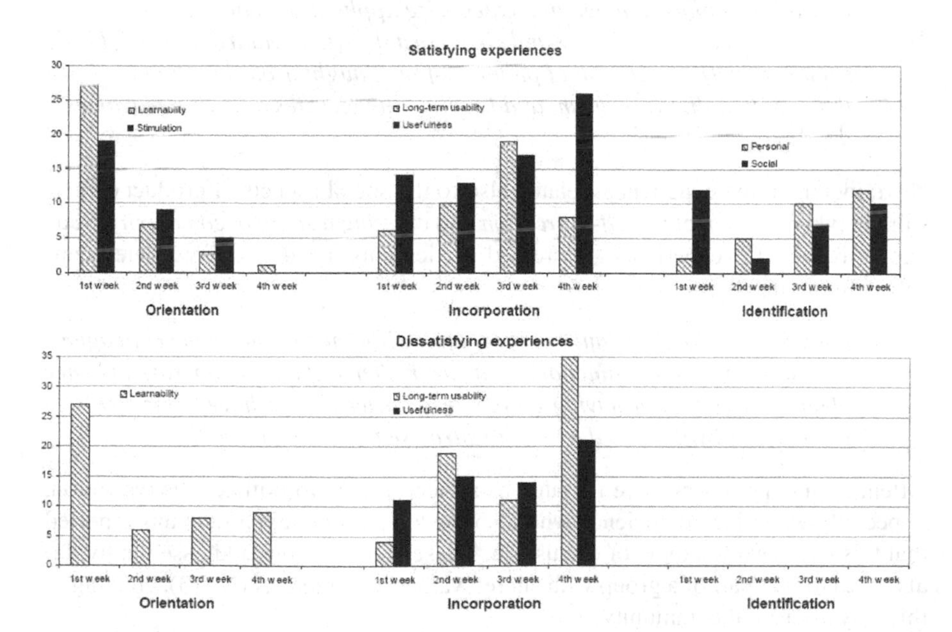

Fig. 4.6 Number of reported satisfying & dissatisfying experiential episodes over the four weeks relating to the three phases of adoption.

Dissatisfying experiences (N=130) related to *long-term usability problems* (N=69), and to *usefulness problems* (N=61), i.e. expected but missing features,

> *"**[Long-term usability problems, day 23]** "When I wear gloves I am not able to work with iPhone. It is really impractical when I am cycling or riding a motorcycle", [day 23] "...carrying iPhone in one hand and then pressing the button at the very bottom to take a picture was quite difficult. It is difficult to balance it", **[usefulness problems, day 3]** "... I could not believe it had no zoom! I messed around for a while but all in vain. Why someone should zoom while taking pictures from iPhone? Right? Simplicity is key...make products simple and do not even give those features which people actually want!!!"*

4.4.3.1.4 Identification

Finally, identification reflected ways in which participants formed a personal relationship with the product as it was increasingly incorporated in their daily routines and interactions.

Identification was found to have two perspectives: *personal* and *social*. Participants were increasingly identifying with the product as they were investing time in *adapting and personalizing* it (N=23), but also as the product was associated with *daily rituals* (N=8):

> ***[personalization, day 14]*** *"I downloaded a new theme ... It looks very beautiful. Now my iPhone looks much much better than before", [day 27] "Today I tried this application again to categorize application icons on the screen... Now my screen looks so nice and clean, just the way I wanted it to be",* ***[daily rituals, day 9]*** *"I put a lot of pictures of my daughter on the iPhone... I like that functionality very much, and I look at the pictures at least a few times a day."*

Next, identification experiences related also to the social aspects of product ownership, in two ways: enabling *self-expression* and *creating a sense of community*. Self-expressive (N=18) experiences addressed participants' need to differentiate themselves from others:

> *[Day 8] "...I had the chance to show off my iPhone to some of my colleagues. I showed them some functions that are rather difficult to operate in other phones... I felt good having a BETTER device. I still have some cards to show which I will in do due time to surprise them even more."*

Often, such experiences were initiated as an ice-breaker to initiate a conversation. Especially when meeting friends who also owned an iPhone, participants reported that this was always a topic of discussion. These conversations addressed individuals' need to feel part of a group with shared values and interests (N=13), creating in this way a sense of community:

> *[Day 25] "Yet another friend of ours has an iPhone. It's a guaranteed subject of conversation if you see another person having it... we chatted about how*

many applications and which we have on it. It is nice to get recommendations for new cool stuff you could use"

Experiences relating to Identification displayed a more complex trend (figure 4.6). While experiences reflecting the personal side of identification increased over time, social experiences displayed an initial decrease, but also a gradual and sustaining increase. These two patterns were found to be rooted in distinct aspects of social identification. Experiences relating to self-expression (median day=8), e.g. announcing the recent possession in social contexts, wore off along with users' initial excitement. Experiences relating to the feeling of being part of a community sharing similar values and interests, however, displayed an increasing and sustaining effect (median day=24).

4.4.3.2 What Motivates the Transition across These Phases?

These temporal patterns were found to relate to three underlying forces: *familiarity*, *functional dependency* and *social and emotional attachment*. First, as users' familiarity with the product increased, the number of experiences relating to learnability problems, but also stimulation and self-expressive identification decreased:

> *[Day 15] "My typing speed on iphone is gradually improving... now I am a big fan of this keyboard and I find it very comfortable and easy to use", [Day 20] "With today's busy schedule I didn't even remember I had an iPhone. I think the toy becomes just a nice daily use item - usable and good to have but the initial excitement seems to be gone."*

Second, as users incorporated the product in their daily lives, they were experiencing an increasing functional dependency, resulting in experiences relating to the product's usefulness and long-term usability:

> *[day 10] "...I am becoming a great fan of it. It's simply saving time", [Day 15] "...I've slowly started adapting to those things and I must say it feels like my phone-life got a little bit easier."*

Last, as the product is incorporated in users' lives, it not only provides the benefits that were intended by the designers but also becomes a personal object, participateing in private and social contexts, resulting in an increasing emotional attachment to the product:

> *[Day 18] "My daughter seems to be attracted to everything that shines, and whenever she spots the iPhone she grabs it. I try to distract her, by giving her the iPhone's case. Unfortunately she is smarter than that : I find it very funny to see that she likes the same things as me", [Day 2] "In the evening we had friends over for dinner. They are also quite technology freaks. Quite quickly I told them that I've got an iPhone and showed it to them. I really liked watching them playing with it..."*

4.4.3.3 How Does Each Phase Contribute to the Overall Perceived Quality of the Product?

Hassenzahl (2004) distinguished between two overall evaluative judgments of the quality of interactive products, namely judgments of Goodness and of Beauty. While prior work suggests goodness to be a goal-oriented evaluation, relating to the pragmatic quality of the product (usefulness and ease-of-use), and beauty a pleasure-oriented evaluation, relating to hedonic quality (stimulation and identification) (Hassenzahl, 2004; Tractinsky and Zmiri, 2006; Mahlke, 2006; Van Schaik and Ling, 2008), we saw something different. In each phase, different qualities of the product were crucial for its gradual acceptance.

Tables 4.4 and 4.5 display regression analyses with usefulness, ease-of-use, stimulation and identification as predictors and Goodness or Beauty as predicted variables. Each variable is measured through a single item. Each case depicts users' ratings within a single experience. Cases were categorized in three groups based on whether an individual experience was classified as relating to the orientation, the incorporation, or the identification phase.

While during *Orientation* the ***Goodness*** of the product was primarily derived on the basis of its ease-of-use (Regression analysis: β=0.43, t=4.79, p<.001) and stimulation (β=0.43, t=4.79, p<.001), in *Incorporation*, the product's usefulness (β=0.49, t=10.84, p<.001) became the primary predictor of Goodness, and in the phase of *Identification* the qualities of identification (β=0.53, t=3.57, p<.01) and ease-of-use (β=0.44, t=2.96, p<.01) became the most dominant qualities impacting the overall goodness of the product.

Table 4.4 Multiple Regression analysis with usefulness, ease-of-use, stimulation and identification as predictors and ***Goodness*** as predicted (β values and significances * p<.05, ** p<.001) for both satisfying and dissatisfying experiences.

	Orientation	***Incorporation***	***Identification***
Usefulness		.49**	
Ease-of-use	.43**	.19**	.44**
Stimulation	.43**	.22**	
Identification		.14**	.53**
Adjusted R^2	.63	.79	.51

As expected, ***Beauty*** appeared to be highly related to the quality of identification, i.e. the social meanings that the product communicates about its owner (Orientation: β=0.51, t=4.32, p<.001, Incorporation: β=0.47, t=8.17, p<.001, Identification: β=0.78, t=5.73, p<.001), and stimulation (Orientation: β=0.22, t=1.89, p=.06, Incorporation: β=0.27, t=4.69, p<.001).

Next, we found a priori expectations to have surprisingly limited impact on the actual experience with the product. Based on earlier research, one would expect a priori expectations to have a major role in forming overall evaluative judgments

Table 4.5 Multiple Regression analysis with usefulness, ease-of-use, stimulation and identification as predictors and ***Beauty*** as predicted (β values and significances * p<.05, ** p<.001) for both satisfying and dissatisfying experiences.

	Orientation	***Incorporation***	***Identification***
Usefulness			
Ease-of-use			
Stimulation	.22*	.27**	
Identification	.51**	.47**	.78**
Adjusted R^2	.47	.44	.59

(Lindgaard and Dudek, 2003). Confirming a priori expectations has been seen as the major source of satisfaction both in HCI (Lindgaard and Dudek, 2003) and Information Systems (Oliver, 1980) research. The comparison standards paradigm (Oliver, 1980), which dominates user satisfaction research, posits that individuals form stable expectations to which the actual product performance is compared, to derive a satisfaction judgment. In this study, we saw a priori expectations to evolve in a number of ways.

For 72% of a priori expectations, participants reported a change in their perceived importance. 19% of participants' expectations exhibited a decrease in their importance over time. Although these expectations were on average disconfirmed (i.e., median=3 on a 7-point scale), they did not lead to dissatisfaction (median=5). This was attributed to two major phenomena: *transition from fantasy to reality*, and *post-purchase situational impact variations*. First, participants reflected that these expectations were unrealistically high, i.e., *"[they] hoped for, but not expected"*. As a result, disconfirmation of these expectations was not attributed to the product as a failure but rather to their own perceptions as a 'loss of illusion'. Second, as users were incorporating the product in their routines, the design space was adapting. For example, some participants became less concerned about the coverage of mobile internet through the cell network as they found themselves having access to internet mostly over WiFi networks, while others became less concerned about the ease with which 3rd party applications are being installed as they found themselves satisfied with the pre-installed ones.

53% of a priori expectations exhibited an increase in their importance over time. The majority of these expectations (87%) were either confirmed or exceeded. The major source of the increase in their perceived importance was participants' initial inability to judge the impact of the expected feature in the long run. As participants incorporated the feature in their daily lives, they were becoming more dependent on it and its perceived importance was increasing. These expectations mostly related to the use of mobile internet, mobile agenda, and to the effectiveness and efficiency of the virtual keyboard.

4.4.4 Discussion

Overall, we showed time to be a significant factor altering the way individuals experience and evaluate products. We identified distinct phases in product adoption and use, which we summarize here.

4.4.4.1 From Orientation to Incorporation

The impact of novelty in users' experience displayed a sharp decrease after the first week of use. Contrary to common belief that iPhone's success is largely due to its aesthetics and novel interaction style, these aspects were found to play a minimal role in providing positive prolonged experiences.

Next, we found a shift in users' concerns over time from ease-of-use to usefulness. While ease-of-use was the primary predictor of goodness during early orientation experiences, usefulness became an even more dominant predictor during the incorporation phase. This resembles recent research in the field of ubiquitous computing urging for a shift in the emphasis from efficient use to meaningful incorporation (Hallnäs and Redström, 2002; Davidoff et al., 2006). Moreover, the types of interaction problems that users experienced shifted over time, in support of the findings of Mendoza and Novick (2005) and Barendregt et al. (2006). While early use was described by learnability problems induced by unexpected product behavior, prolonged usability related to repeating problems, often rooted in unanticipated use.

4.4.4.2 From Incorporation to Identification

Participants were found to develop an emotional attachment to the product as they increasingly incorporated it in their daily life. We found emotional attachment to be closely tied to the type of product. The iPhone is a very personal product as it connects users to loved persons, allows adaptation to personal preferences, and is always nearby. It is also a very social product as it communicates qualities of self-identity and connects to others by displaying shared values and interests. It is unknown how emotional attachment will develop with products that do not participate in users' personal and social interactions.

4.4.4.3 Actual Experience More Influential Than Expectations

While earlier work (Lindgaard and Dudek, 2003) would suggest that a priori expectations play a major role in the formation of satisfaction judgments, we found them to evolve during the actual experience with the product. Often, this was induced by lack of knowledge. As users' experience with certain features exceeded their a priori expectations, these features became increasingly important to overall satisfaction with the product.

At the same time, disconfirmed expectations seemed to become less relevant to users' satisfaction over time. A possible explanation for this could be supported by the theory of Cognitive Dissonance (Festinger, 1957), which postulates that after a purchase there is a certain degree of psychological discomfort rooted in the discrepancy between the desired and the actual outcome of the choice. The most likely outcome of dissonance is attitude spread, namely, participant's efforts in justifying their choice by adapting their a priori expected outcome, or in our context, the perceived importance of their expectations.

All in all, the actual experience with the product seemed to be more influential to participants' satisfaction judgments than their a priori expectations. Note, that we do not claim that forming expectations about a future possession does not influence experience; instead, we believe the act of anticipation to be a crucial part of our experience. Often, anticipating our experiences with a product, becomes even more important, emotional, and memorable than the experiences per se. It is only when conflicting with actual experience that a priori expectations appear to adapt in an effort of reducing the discrepancy between expected and actual experience.

Finally, what makes a product good and beautiful? Most studies suggest that goodness is a goal-oriented evaluation, related to pragmatic quality perceptions and beauty a pleasure-oriented evaluation related to hedonic quality perceptions (Hassenzahl, 2004; Mahlke, 2006; Tractinsky and Zmiri, 2006; Van Schaik and Ling, 2008).

The current study has diversified this view. While *goodness* was on average related to pragmatic quality perceptions, it was significantly affected by stimulation during orientation and by identification during social experiences. These findings suggest that the overall value, or the goodness of a product is contextually dependent, e.g. a novel product will be better than a more traditional one during our initial interactions but not necessarily in our prolonged experiences. Overall, we showed time to be a significant factor altering the way individuals experience and evaluate products.

4.4.5 Implications for Design

What does this work suggest to HCI practice? HCI has up to now mostly focused on early interactions. As a consequence we have been mostly concerned about the product qualities that dominate in early use. We argue that the focus of HCI practice should expand from the study of early interactions to the study of prolonged experiences, understanding how a product becomes meaningful in a person's life. We therefore promote three interesting avenues for further research.

4.4.5.1 Designing for Meaningful Mediation

What contributes to the successful appropriation of products? When does a product become useful in one's life? We found usefulness to be much broader than the functionality of the product, relating to the impact of the functionality in participants' lives. iPhone's usefulness emerged through its appropriation in specific contexts and the changes this brought to participants' lives. For instance, the reflection of one of the participants on the Notes^{TM} functionality was that it provided the freedom of going for jogging whenever she wanted to think of her work, as she could easily write down notes while being mobile (c.f. "capturing momentary information"). Usefulness, in this case, was derived from supporting her need for autonomy, being able to combine physical exercise and progress in her work.

On one hand, this provides hints that the product's usefulness emerges in a process of appropriation in certain contexts of use, and thus may not become evident in early use and user tests involving minimal exposure to the product. On the other hand, one could speculate that this context of use was most likely not anticipated during the design of the iPhone. The question raised then is, how can we design for contexts that we cannot anticipate? We believe iPhone's success here to be rooted in what Taylor and Swan (2005) call designing for artful appropriation, i.e. designs that are specific enough to address one single need, but flexible enough to enable the artful appropriation in diverse contexts.

4.4.5.2 Designing for Daily Rituals

People love parts of their daily lives and the products that are associated with them. Drinking a cup of coffee after waking up, listening to one's favorite songs while driving home, drinking a glass of wine in the evening; these are some examples of activities that become habituated and cherished. We found activities mediated through the iPhone, like checking for new emails after waking up, or looking at a daughter's photos several times during the day gradually becoming daily rituals that people love to perform. But, how can we design for new daily rituals? How can we identify the activities that people love in their daily lives if these are habituated and perhaps not apparent to the individual? It is crucial to follow the appropriation of products in participants' lives, but also to understand the impact of the forfeiture of these products once these have been embedded in habituated activities.

4.4.5.3 Designing for the Self

People become attached to products that support a self-identity they desire to communicate in certain settings (Belk, 1988). The iPhone supported two needs in participants' social experiences: self-expression and differentiation from others (e.g. showing off to friends and colleagues), as well as a need for integration and feeling part of a group.

Products and self-identity have been a major part of consumer behavior research, but remain largely unexplored in HCI and design research. How can we understand the social meanings that users communicate through the possession of products? And how can we adapt our task-focused HCI methods to design for the more experiential aspects of product use and ownership like the social meanings of products? One example could be the work of Zimmerman (2009) who proposes techniques for understanding and designing for the dynamics of self-identity where individuals have to reinvent themselves in a new role.

4.5 Discussion

This chapter has discussed two studies that tried to assess how users' experiences with products develop over time. Besides having distinct goals, they employed diverse methodology for *"measuring"* the user experience: a *reductionist* and a *holistic* one.

The first study was reductionist in nature. It employed a validated measurement model (Hassenzahl, 2004) and sampled users perceptions across two points in time. It then tried to identify variations in the structural relationships between the latent constructs, i.e. the individual quality perceptions and the overall judgments of goodness and beauty. While longitudinal studies in user experience are scarce, similar methodological approaches can be found in the field of Technology Acceptance (Venkatesh and Davis, 2000; Venkatesh and Johnson, 2002; Kim and Malhotra, 2005). Such studies typically employ validated structural models across different phases in the adoption of a system. For instance, Venkatesh and Davis (2000) employed the Technology Acceptance Model (Davis et al., 1989) over three points in the adoption of information systems at work settings: before the introduction of the system (inquiring into users' expectations), right after the introduction of the system, and three months after the introduction.

An assumption inherent in this approach is that the relevant latent constructs remain constant, but their perceived value and relative dominance might alter over time. But, especially in developing fields such as that of user experience, substantial variations might occur over time even in what constructs are relevant to measure. Some constructs, e.g. novelty, might cease to be relevant while others that were not evident in studies of initial use might become critical for the long-term acceptance of a product. Note for instance, the wider spectrum of experiences relating to *daily rituals* and *personalization* that could not be captured by the measurement model that we employed in the first study. This challenges the content validity of the measurement model as relevant latent constructs might be omitted, but may also lead to distorted data as the participant might fail in interpreting the personal relevance of a given scale item to her own context, for instance when a latent construct and its individual scale items cease to be relevant. In such cases participants may shallowly process the statement of the scale and ratings may reflect superficial language features of the scales rather than participant's perceptions (Larsen et al., 2008b).

Moreover, such approaches provide rather limited insight into the exact reasons for changes in users' experiences. They may, for instance, reveal a shift in the dominance of perceived ease-of-use and perceived usefulness on intention to use a product (e.g. Venkatesh and Davis, 2000), but provide limited insight into the exact experiences that contributed to such changes, the underlying motivations for changes in users' preferences, and the contextual variations in product use.

Study 2 was more effective in providing rich insights into the exact reasons for the dynamics of users' experiences over time. Beyond eliciting anecdotal reports on users' experiences, through content analysis of the narratives, we were able to quantify the dominance of different product qualities on users' overall evaluative judgments. It enabled capturing aspects of experience that were not identified a-priori, but also quantifying their significance and understanding how these relations changed over time.

In this view, experience narratives provide at least two kinds of information. Firstly, each narrative provides rich insights into a given experience and the context in which it takes place. Secondly, generalized knowledge may also be gained from these experience narratives. Such generalized knowledge may be reflected in questions like: how frequent is a certain kind of experience, what is the ratio of positive versus negative experiences and how does this compare to competitive products, how does the dominance of different product qualities fluctuate over time and what should we improve to motivate prolonged use?

This leads to two research problems. Firstly, how can we elicit experience narratives efficiently? Longitudinal designs such as the one employed in study 2 are labor intensive and, consequently, they are often restricted in terms of user population, product population, and studied time. In chapter 5 we review existing methodological approaches for studying temporal changes in user experience and present a novel survey technique that aims at assisting users in self-reporting their most impactful experiences with a product.

Secondly, how can we aggregate the idiosyncratic experiences into generalized knowledge? In chapter 6 we propose a novel technique for the semi-automatic analysis of experience narratives that combines traditional qualitative coding procedures (Strauss and Corbin, 1998) with computational approaches for assessing the semantic similarity between documents (Salton et al., 1975).

4.6 Conclusion

This chapter presented two studies that aimed to assess the differences between initial and prolonged experiences with interactive products.

The first study assessed the ways in which 10 individuals formed overall evaluations of a novel pointing device across two points in the adoption of the product: during the first week and after four weeks of use. Findings suggested judgments of the overall *goodness* of the product to shift from a use-based evaluation dominated by the pragmatic quality of the product, i.e. usefulness and ease-of-use, to an ownership-based evaluation dominated by aspects of identification, i.e. what the product expressed about their self-identify in social contexts. Judgments of *beauty* seemed to be dominated by perceptions of novelty during initial interactions, but this effect seemed to disappear after four weeks of use.

The second study followed six individuals through an actual purchase of the Apple iPhone and inquired into how their expectations and experiences developed one week before and four weeks after the purchase of the product. The study revealed that the product qualities that provided positive initial experiences were not very crucial for motivating prolonged use. Product adoption contained three distinct phases: an initial orientation to the product dominated by the qualities of stimulation and learnability, a subsequent incorporation of the product in daily routines where usefulness and long-term usability became more important, and finally, a phase of increased identification with the product as it participated in users' personal and social experiences. We conceptualized temporality of experience as consisting of three main forces, an increasing familiarity, functional dependency and emotional attachment, all responsible for shifting users experiences across the three phases in the adoption of the product. Based on the findings, we promoted three directions for future HCI practice: designing for meaningful mediation, designing for daily rituals, and designing for the self.

Next to providing empirical insights into the dynamics of experience over time, these two studies raised a number of methodological issues in the study of time in the user experience field. we highlighted a number of limitations of traditional reductionist approaches where a-priori defined measurement models are employed in measuring the user experience. We suggests an alternative methodological approach that relies on a) eliciting the experiences that are personally meaningful to each participant, in the form of *experience narratives*, and b) employing content analysis techniques in creating multiple levels of abstraction, from concrete idiosyncratic insights to abstracted and generalized knowledge. We concluded by raising two research questions that will be addressed in the chapters and respectively.

Chapter 5
iScale: Studying Long-Term Experiences through Memory

Abstract. This chapter presents iScale, a survey tool for the retrospective elicitation of longitudinal user experience data. iScale aims to minimize retrospection bias and employs graphing to impose a process during the reconstruction of one's experiences. Two versions, the *constructive* and the *value-account* iScale, were motivated by two distinct theories on how people reconstruct emotional experiences from memory. These two versions were tested in two separate studies. Study 1 aimed at providing qualitative insight into the use of iScale and compared its performance to that of free-hand graphing. Study 2 compared the two versions of iScale to free recall, a control condition that does not impose structure on the reconstruction process. Overall, iScale resulted in an increase in the amount, the richness, and the test-retest consistency of recalled information as compared to free recall. These results provide support for the viability of retrospective techniques as a cost-effective alternative to longitudinal studies.

5.1 Introduction

Understanding the use and acceptance of interactive products beyond initial use has always been an interest of the Human-Computer Interaction (HCI) community (Erickson, 1996; Prümper et al., 1992). However, two recent trends make the call for a more longitudinal view more urgent (Karapanos et al., 2009c). First, legislation and competition within the consumer electronics industry have led to prolonged product warranties, resulting in an alarmingly increasing number of products being returned on the basis of failing to satisfy their users' "true" needs (Den Ouden et al., 2006). Second, products have become more embedded into services. Often, products are being sold for low prices or even given away for free and revenues stem mainly from the supported service and their prolonged use (Karapanos et al., 2009c). Thus, the overall focus on product quality shifts from a focus on the classic phases of pre-purchase and purchase to a more longitudinal perspective, trying to better understand use and liking over time. This is a shift increasingly taken up by the HCI

E. Karapanos: Modeling Users' Experiences with Interact. Syst., SCI 436, pp. 85–114.
springerlink.com

community (Gerken et al., 2007; Barendregt et al., 2006; Fenko et al., 2009; Karapanos et al., 2008a; von Wilamowitz Moellendorff et al., 2006; Courage et al., 2009; Vaughan et al., 2008; Kjeldskov et al., 2008).

From a methodological perspective, one may distinguish three approaches to understanding the development of usage and experience over time (von Wilamowitz Moellendorff et al., 2006): cross-sectional, pre-post/longitudinal, and retrospective reconstruction. Cross-sectional approaches are the most popular in the HCI domain (Prümper et al., 1992; Bednarik et al., 2005). Cross-sectional studies distinguish, for example, user groups with different levels of expertise, for instance, novice and expert users. Observed variation in experience or behavior is then attributed to expertise in the sense of a quasi-experimental variable. This approach is, however, limited as it is prone to confounding variables, such as failing to control for external variation and, more importantly, falsely attributing variation across the user groups to expertise. Prümper et al. (1992) already highlighted this problem, by showing that different definitions of novice and expert users lead to different results.

Beyond the cross-sectional, one may further distinguish pre-post and true longitudinal approaches. Pre-post designs study the same participants at two points in time. For instance, Kjeldskov et al. (2008) studied the same seven nurses, using a healthcare system, right after the system was introduced and 15 months later. Karapanos et al. (2008a) explored how ten individuals formed overall evaluative judgments of a novel pointing device, during the first week of use as well as after four weeks of using the product. While these approaches study the same participants over an extended period of time, they cannot tell much about the exact form of change, due to the limited number of only two observations. True longitudinal designs take more measurements and employ a number of statistical techniques to track change in general and to estimate the impact of particular events on change. Because of their laborious nature, however, they are only rarely used in practice and research.

Different granularities in longitudinal studies can be distinguished (see von Wilamowitz Moellendorff et al., 2006): a micro perspective (e.g. an hour), a meso perspective (e.g. 5 weeks) and a macro perspective, with a scope of years of use. Studies with a micro-perspective assess how users' experience changes through increased exposure over the course of a single session of use. For instance, Minge (2008) elicited judgments of perceived usability, innovativeness and the overall attractiveness of computer-based simulations of a digital audio player at three distinct points: a) after participants had seen but not interacted with the product, b) after two minutes of interaction and c) after 15 minutes of interaction. An example of a study with a meso-perspective is Karapanos et al. (2009c). They followed six individuals after the purchase of a product over the course of 5 weeks. One week before the purchase of the product, participants started reporting their expectations. After product purchase, participants were asked to narrate the three most influential experiences of each day. Studies with a macro-perspective are 'nearly non-existent' (von Wilamowitz Moellendorff et al., 2006).

A third approach is the retrospective reconstruction of personally meaningful experiences from memory. Different variants of the Critical Incident Technique, popular in marketing and service management research (Edvardsson and Roos, 2001;

Flanagan, 1954), ask participants to report critical incidents over periods of weeks, months or the complete time-span of the use of a product or service. In a survey study, Fenko et al. (2009) asked participants to recall their single most pleasant and unpleasant experience with different types of products and to assess the most important sensory modality (i.e. vision, audition, touch, smell and taste) at different points in time (i.e. when choosing the product in the shop, during the first week, after the first month, and after the first year of usage). Von Wilamowitz-Moellendorff et al. (2006; 2007) proposed a structured interview technique called CORPUS (Change Oriented analysis of the Relation between Product and User) for the retrospective assessment of the dynamics in users' perceptions of different facets of perceived product quality. CORPUS starts by asking participants to assess the currently perceived quality of "their" product on a number of defined facets (usability, utility, beauty, stimulation, identification, and global evaluation). Subsequently, they are asked to "go back in time" and to compare their current perception and evaluation of the product to the moment right after purchasing the product. If change has occurred, participants are further prompted to indicate the direction and shape of change (e.g., accelerated improvement, steady deterioration). Finally, participants are asked to elaborate the reasons that induced the changes in the form of short narratives, so-called "change incidents". The obtained data can be used quantitatively by constructing graphs of change (see Figure 5.1 for an example) and qualitatively by exploring the reasons people give for changes.

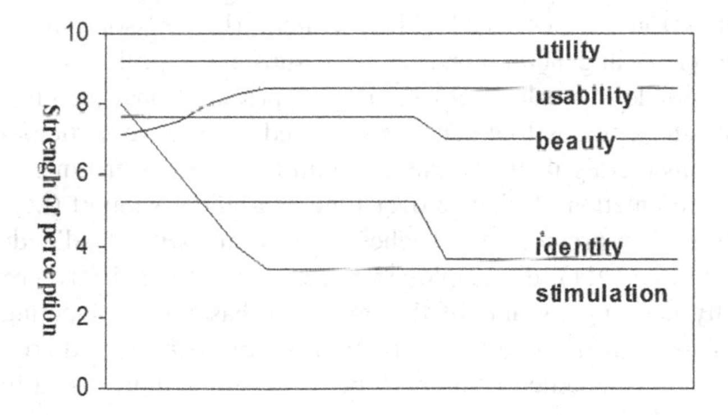

Fig. 5.1 Exemplary dynamics of different perceived quality dimension of mobile phones. Start and end point of each dimension are based on the mean elicited ratings of eight participants. Reprinted from von Wilamowitz Moellendorff et al. (2006)

A common critique of methods relying on memory is the degree to which recalled experiences are biased or incomplete. In the context of perceived product quality, we argue that this is of minor importance. While a given reconstruction from memory should be truthful (i.e., reflect what the participant really thinks), it seems less important, whether the reconstructed timeline as well as the reasons given are true (i.e. reflect what actually happened) as long as the participant is convinced that what she is reporting actually happened. This is because, we are foremost interested in subjective reconstructions because those (and not "objective" data) will be communicated to others as well as guide the individual's future activities. In other words, it may not matter how good a product is objectively, it is the "subjective", the "experienced", which matters (Hassenzahl et al., 2006). See also Norman (2009). To give a further example: Redelmeier and Kahneman (1996) found retrospective assessment of the pain experienced during colonoscopy to be biased. People put an extra weight on the most painful moment and the end of the examination. This has interesting consequences. One can, for example, deliberately prolong the examination (something not approved by the patients), but make sure that these last, additional two minutes are not painful. The consequence is an overall assessment of the examination as less painful compared to people without the additional two minutes. While this is clearly a bias, people simply have no memory for all the moments they experience, but will remember their overall impression of the examination. The retrospective judgment is more real to them than what actually happened. While the validity of remembered experiences may not be crucial, their consistency across multiple recalls is. It seems at least desirable that participants would report their experiences consistently over multiple trials. If recall would be purely "random", the value of respective reports for design would be questionable. In other words, what we remember might be different from what we experienced; however, as long as these memories are consistent over multiple recalls, they provide valuable information.

In the area of critical incident research, interviewing techniques have been developed with the aim to assist participants in remembering more details of and contextual information around experienced critical incidents (Edvardsson and Roos, 2001). However, interviews in general, however, need substantial skills and resources. It, thus, seems desirable to create a self-reporting approach. Consequently, this chapter presents iScale, a survey tool that was designed to increase participants' effectiveness in reconstructing their experiences with a product over time. iScale uses a graphical representation of change over time as a major support (i.e., time-line graphing). Other than previous approaches (von Wilamowitz Moellendorff et al., 2006; Kujala et al., 2011), the employed procedure is more firmly grounded on theory, actually deriving variants of the procedure based on competing theoretical models of the retrospective reconstruction of episodes and experiences from memory. Graphing is assumed to support the reconstruction process through what Goldschmidt (1991) calls *interactive imagery* (i.e., "the simultaneous or almost simultaneous production of a display and the generation of an image that it triggers"). The idea of using graphing as an approach to introspecting on past emotional experiences can be traced back to Sonnemans and Frijda (1994).

We begin with laying out two different ways of obtaining retrospective reconstructions of experiences and their theoretical foundation. We then present the results of two studies. Study 1 acquired a qualitative understanding of the use of iScale in comparison to its analog equivalent (i.e. free-hand graphing). Study 2 assessed how iScale compares to an experience reporting tool without graphic support, which, can be seen as a control condition to assess the impact of iScale on participants' effectiveness and test-retest consistency in reconstructing experiences.

5.2 Reconstructing Experiences from Memory

Memory was for long understood as a faithful account of past events, which can be reproduced, when trying to remember details of the past. This idea was first challenged by Bartlett (1932). He described remembering as an act of reconstruction, which never produces the exact past event, but instead alters representation of the event with every attempt to recall. Bartlett (1932) asked participants to recall an unfamiliar story told 20 hours earlier. The recalled stories differed from the original in detail, order and importance of single events. In addition, participants augmented their memories by applying rationalizations and interpretations to the original story. Each further reconstruction distorted the stories even further.

At the heart of the notion of reconstruction lies the distinction between episodic and semantic memory (Tulving, 2002). While episodic memory "is specific to a particular event from the past, semantic memory is not tied to any particular event but rather consists of certain generalizations (i.e. beliefs) that are rarely updated" (Robinson and Clore, 2002, p. 935). These two types of memory serve different needs, such as learning new information quickly - a capacity of episodic memory - or developing relatively stable expectations about the world - a capacity of semantic memory (Robinson and Clore, 2002). Reconstruction happens through the retrieval of cues stored in episodic memory. In the absence of such cues, beliefs stored in semantic memory may be used to reconstruct the past. This results in distortions, where details that actually happened are replaced by generalizations based on what we know about the world. Thus, the accuracy of remembered events depends on the degree to which contextual cues are present and active in episodic memory.

Experiences do not only consist of contextual details, but also of value-charged elements, such as emotions or overall evaluative judgments. One can distinguish two approaches to the reconstruction of value-charged experiences: the *constructive* and the *value-account* approach. The constructive approach assumes that felt emotion cannot be stored in memory but is instead reconstructed from recalled contextual cues. In contrast, the value-account approach proposes the existence of a memory structure, which stores the frequency and intensity of a person's responses to a stimulus. This in turn can be used to cue the recall of contextual details of one's experiences. In the following, we describe the two approaches in more detail.

5.2.1 The Constructive Approach

The constructive approach assumes that reconstruction happens in a forward temporal order (Anderson and Conway, 1993; Barsalou, 1988; Means et al., 1989). Barsalou (1988) asked people to recall their experiences during the last summer. Most participants started in the beginning and proceeded in a chronological order. Often, recalling one episode cues the reconstruction of further episodes and more contextual information surrounding the episode (Anderson and Conway, 1993) - just like a string of pearls.

Robinson and Clore (2002) further argued that "emotional experience can neither be stored nor retrieved" (p. 935), but can only be reconstructed on the basis of recalled contextual cues. They propose an accessibility model that distinguishes between four types of knowledge used to (re)construct an emotion. First, *experiential knowledge* is used when an emotion is constructed "online" (i.e. as the experience takes place). When experiential knowledge is inaccessible, people will resort to *episodic information*: they will recall contextual cues from episodic memory to reconstruct the emotional experience. When episodic memories become inaccessible, people will shift to semantic memory. They first access *situation-specific beliefs*: beliefs "about the emotions that are likely to be elicited in a particular type of situation" (p. 935). If event-specific beliefs are inaccessible as well (e.g., due to rarity of the event) people will access *identity-related beliefs*: "beliefs about their emotions in general" (p. 935).

Motivated by the accessibility model of Robinson and Clore (2002), Daniel Kahneman and colleagues (2004; 2008) developed the Day Reconstruction Method (DRM), an online diary method that attempts to minimize retrospection biases when recalling emotional experiences. DRM starts by asking participants to mentally reconstruct their daily experiences as a continuous series of episodes, writing a brief name for each one. This aims at eliciting contextual cues within each experiential episode as well as temporal relations between episodes. As a result, participants reconstruct the emotional experience on the basis of a more complete set of episodic information, thereby minimizing bias from overly relying on semantic information, which is detached from the actual experience.

5.2.2 The Value-Account Approach

The value-account approach assumes that people recall an overall emotional assessment of an experience, but not the exact details of the experienced event. Betsch et al. (2001) proposed the existence of a memory structure called *value-account*, which stores the frequency and intensity of positive or negative responses to stimuli. Since Value-Account is more accessible and better retained over time than details from episodic memory, it will actually cue episodic information, when reconstructing an experience, or feeds into a potential overall evaluation even in the absence of any episodic information (Koriat et al., 2000; Neisser, 1981). This top-down process for reconstructing memories is consistent with research on autobiographical memory, where three levels are distinguished: lifetime periods, general events, and event-specific knowledge. Reconstruction takes place in a top-down fashion where knowledge stored at the level of a lifetime period cues information at the two lower levels (Conway and Pleydell-Pearce, 2000).

Both approaches, constructive and value-account, suggest specific processes for retrieving emotional experiences from memory. While the constructive approach assumes a bottom-up, chronological recall of episodic information that cues the reconstruction of experienced affect and emotion, the value-account approach suggests a top-down process, starting from affect stored in specific value-accounts to cue the recall of specific episodic information. In the following section, we will illustrate how these two processes were operationalized in two separate versions of the iScale tool.

5.2.3 *Graphing Affect as a Way to Support the Reconstruction of Experiences*

Imagine being asked to provide a graphical representation of how the global evaluation of your mobile phone changed over time on a timeline, which starts at the moment of purchase and ends in the present. There are a number of ways doing this. The aforementioned CORPUS technique - somewhat atheoretically - recommended starting from the momentary perception, comparing it to the beginning of product use and then recalling specific experiences. The presented theoretical positions above, however, provide clearer, more theoretically grounded suggestions of how to proceed. The constructive approach suggests starting from the beginning, recalling a piece of experiential information (e.g., an episode, a situation) and reconstructing an affective response and the according evaluative judgment (see e.g., Hassenzahl and Ullrich (2007) for the relation between affect and product evaluation) to this piece of information. This will then cue the next piece and so forth until change over time is reconstructed. The value-account, however, suggests starting with the affective information, that is, recalling the change of affect and evaluation over time first, and using the general shape of this to recall more specific experiential information.

Obviously, both approaches can be supported by a tool. However, using the one or the other procedure may impact the results. To study potential differences, we created two different versions of iScale, the constructive and the value-account iScale.

5.2.4 *iScale*

Experience reconstruction with iScale starts with two questions: a) "What was your opinion about the product's [particular quality] just before you purchased it?", and b) "How did your opinion about the product's [particular quality] change since then?" The participants are then presented with a timeline that starts at the moment of purchase and ends in the presence. In general, participants graph linear segments that represent an increase or decrease in their perception and evaluation over a certain period. Each line segment is associated with an identifier, displayed below the segment (figure 5.2a). A participant can click on the segment to report one or more experienced events, which are perceived as a cause of change. For each experience report, the participant can provide a brief name (identifier), a more elaborate description of the experienced event (a narrative), and respond to a number of event-specific questions (figure 5.3a). For the goals of the present studies, we asked participants to recall a) the exact time that the event took place, b) the impact of the event on the participant's overall perception and evaluation, and c) the participant's confidence on the exact details of the report. However, these questions can be adapted to particular research interests.

The two versions of iScale differ in the way graphing was used to support the reconstruction of experiences from memory. More specifically they differed in the progression of graphing (feed-forward versus top-down), and, the existence or absence of concurrency between graphing and reporting.

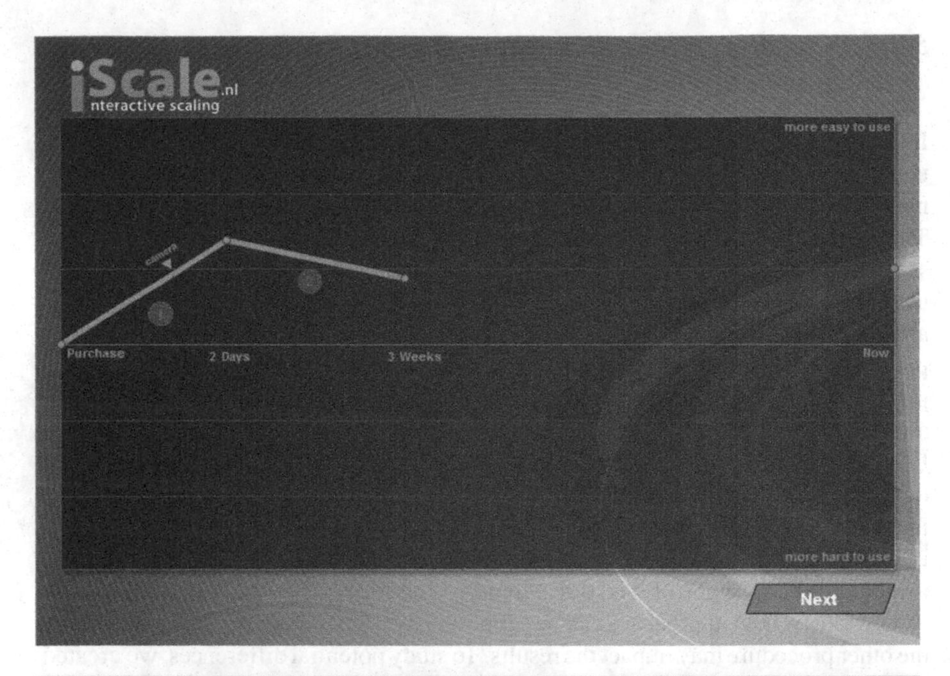

Fig. 5.2 In the *constructive* iScale (top), one starts by plotting points in a serial order (feed-forward progression) and participants are asked to report details of experiences right after a line segment was added (concurrent reporting). In *Value-Account* iScale (bottom), a line connects the start with the end and participants may further split this in distinct periods (top-down progression). Once participants have graphed the full pattern of change over time, they are asked to report on one or more experiences for each line segment (non-concurrent reporting)

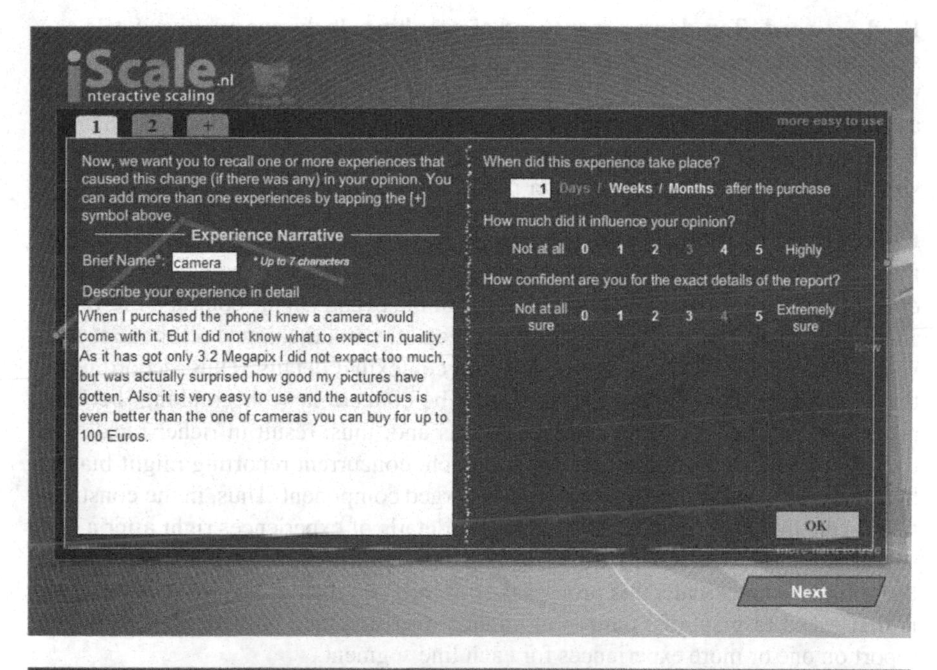

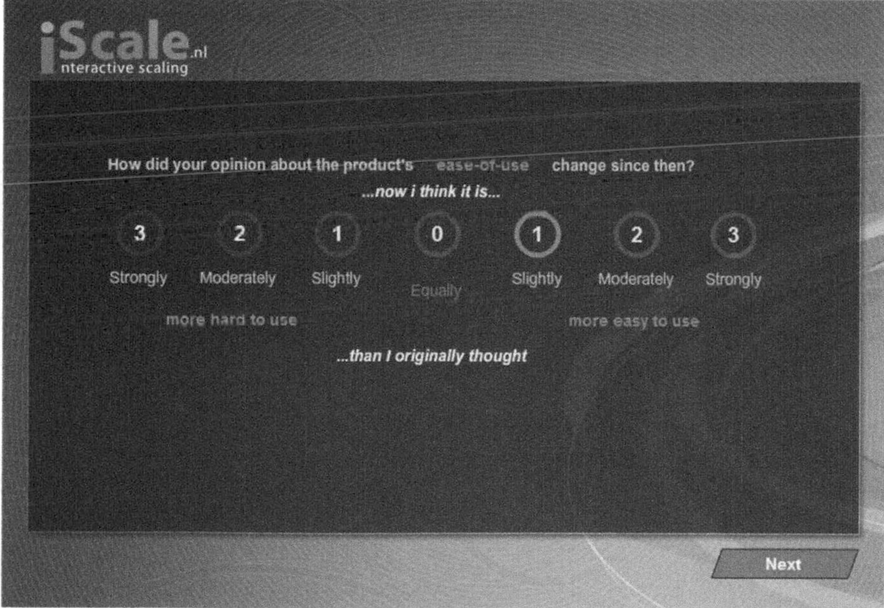

Fig. 5.3 (a) interface for reporting experiences, and (b) overall questions asked in the beginning of the survey

Feed-forward - Top-down progression of graphing: In the constructive iScale one starts by plotting points in a serial order; in the value-account iScale a line connects the start with the end of the timeline using the participant's response to the question asked in the first step about how product perception and evaluation has changed from the moment of purchase to the present. The participant is asked to proceed by splitting the full segment into smaller parts.

Concurrent - Non-concurrent reporting: The constructive approach assumes that the affective, the value-charged component of a past experience can only be reconstructed from recalled contextual details of the event. On the contrary, the value-account approach assumes that individuals recall the value-charged component even without being able to recall the underlying contextual details. Thus, according to the constructive approach reporting should be concurrent with graphing, because reporting will increase the contextual details and, thus, result in richer recall. On the other hand, in the value-account approach, concurrent reporting might bias or hinder the process of recalling the value-charged component. Thus, in the constructive iScale the participant is asked to report details of experiences right after a line segment was added (i.e., drawn). Graphing and reporting proceeds concurrently. In the value-account iScale, this process is split into two distinct steps: the participant is first asked to graph the pattern of change over time and only after that she can report on one or more experiences for each line segment.

In the remainder of the chapter we describe two studies. Study 1 aimed at providing qualitative insight into the use of iScale and compared its performance to free-hand graphing. Study 2 compared the two versions of iScale to free recall, a control condition that does not impose structure on the reconstruction process. Overall, graphing is expected to provide temporal context for the recall of experienced events. This is expected to increase the amount and test-retest consistency of the information that the participants are able to recall. This assumption will be tested in study 2.

5.3 Study 1: Understanding Graphing as a Tool for the Reconstruction of Experiences

The first study aims at a qualitative understanding of graphing as a support for the reconstruction of experiences. It compares the two iScale tools to free-hand graphing.

5.3.1 Method

5.3.1.1 Participants

Twelve graduate students in HCI (7 male, median age 30 years) participated in the study. They were selected based on the diversity in their educational backgrounds.

They were: five Computer Scientists, three Industrial Engineers, two Linguists, one Psychologist and one Industrial Designer.

5.3.1.2 Procedure

The study consisted of two parts. In the first part, each participant used three different graphing techniques: free-hand graphing (FHG) and the two iScale tools (constructive & value-account). All tasks were carried out on a Wacom Cintiq 21UX Interactive Pen Display. The order in which the tools were employed was counterbalanced across participants; FHG was always used first to avoid any bias from the iScale tools as we wished to understand users' natural behavior in graphing changes in perception, evaluation and feelings over time.

Participants were asked to graph how their opinion on three distinct qualities of their mobile phone changed over time (see table 5.1). Each quality was described by a brief definition and three attributes to support the definition (Hassenzahl, 2004; von Wilamowitz Moellendorff et al., 2006). Qualities along with their definitions were derived from von Wilamowitz Moellendorff et al. (2006), though we rephrased the construct names utility and stimulation to usefulness and innovativeness as the former were not clear to some participants in our pilot tests. These qualities are routed in Hassenzahl's (2004) model that distinguishes *pragmatic quality*, which refers to the product's ability to support the achievement of do-goals such as making a telephone call, from *hedonic quality*, which refers to the product's ability to support the achievement of be-goals such as being stimulated or being admired.

Participants were instructed to think aloud; interactions and verbal data were captured on video. Each graphing task took approximately four minutes (min=81 sec, max=594 sec). No significant differences were found between the three tools (constructive: M=230 sec, SD=147 sec; value-account: M=205 sec, SD=87 sec; free-hand graphing: M=301 sec, SD=130 sec).

> *"While graphing, you are asked to report experiences and events that induced these changes in your view of the product. We are interested in your exact thoughts and feelings as you perform the graphing. Why do you graph it in this particular way? What details of events, incidents, experiences do you remember? Is it just a feeling? Please think aloud while doing this."*

Table 5.1 The three aspects, their definition and attributes.

Name	Definition	Word items
Usefulness	The ability of a product to provide the necessary functions for given tasks.	Useful, Practical, Meaningful
Ease-of-use	The ability of a product to provide the functions in an easy and efficient way.	Easy to use, Simple, Clear
Innovativeness	The ability of a product to excite the user through its novelty.	Innovative, Exciting, Creative

In the second part, participants were interviewed about the differences between the three graphing techniques, using a structured interview technique, called the Repertory Grid (Fransella et al., 2003). This technique is well aligned with parallel design, and particularly in the existence of three or more alternative artifacts, and allows for inquiring into participants' idiosyncratic ways in which they differentiate the artifacts. In this way, one can inquire into the design space from a users' perspective (Hassenzahl and Wessler, 2000). This is useful not only for designed artifacts, but also for methods as in the present case. Participants were given three cards, each providing a name and a screenshot of one of the three graphing techniques. Participants were first asked to identify the three techniques. Next, they were asked to "think of a property or quality that makes two of the graphing techniques alike and discriminates them from the third". They were instructed to feel free to make any combination of the three alternatives. Contrary to common practice with the Repertory Grid Technique, we did not probe participants for the exact opposite of the property they provided, but rather focused on further elaboration, when possible. This was supported by *laddering* and *pyramiding* techniques (Reynolds and Gutman, 1988). Laddering seeks to understand what "motivates" a given property and thus ladders up in an assumed means-ends-chain (Gutman, 1982) towards more abstract properties of the stimuli; in laddering we first asked the participant whether the mentioned property is positive, and subsequently why this property is important to him/her (e.g. "why is expressiveness important to you?"). Pyramiding, also known as negative laddering, seeks to understand the lower level properties that make up for a given property; in pyramiding we asked the participant to elaborate on what makes the given technique to be characterized with the respective property (e.g. "what makes free-hand graphing more expressive?").

5.3.2 *Analysis and Results*

5.3.2.1 Understanding Free-Hand Graphing

We analysed FHG to get an idea of how people actually graph changes in product perception and evaluation. We segmented the collected graphs in discrete units. A unit was coded when two conditions were met: a semantic change in the participant's verbal report following a pause in graphing as observed in the video recorded sessions (e.g., "[pause in graphing] but then I got to the point where I got new software updates"). Pauses often suggested an initiation of a new recall. Often this was combined with a change in the slope of the curve, but this was not always the case.

Each unit was then coded for the type of curve and the type of verbal report. Curves were classified into four categories: a) Constant (C) signifying no change in participant's opinion over a certain period, b) Linear (L), either Increasing or Decreasing, c) Non-linear (NL) when there were no grounds for arguing that the curve could be approximated by a linear one or when a single report was associated with two discrete linear curves of different slope (see 5.4b), and d) Discontinuous (D) when the slope was approximately parallel to the vertical axis.

Table 2, line Overall, shows the distribution of the four different types of curves. The majority of segments (44 of 74, 60%) were categorized as linear. Only 5% (4 of 74) of segments were non-linear. Of those, only a single report was associated with two or more linear segments with different slopes (cf. figure 5.4a segment 2, figure 5.4b segment 6, figure 5.4d, segment 1). In addition, only 4 of 74 (5%) instances of discontinuity were observed in the graphs. Thus, while in some cases users are inclined to draw non-linear curves, the majority of curves were linear. This allowed us to focus iScale on linear graphing, thereby reducing a number of potential problems with handling complex, non-linear types of curves in an online tool.

Table 5.2 Relationship between graphing and reporting in free-hand graphing. Types of graphing: C=Constant, L=Linear, NL=Non-Linear and D=Discontinuous

	Type of graph				
Type of report	C	L	NL	D	
Discrete experience	3	30	2	2	37 (50%)
Overall evaluation without motivation	17	4	1	2	24 (32%)
Overall evaluation, motivated by experience	2	10	1	0	13 (18%)
Overall	22 (30%)	44 (60%)	4 (5%)	4 (5%)	74

To get an idea of how the graphs relate to reported detailed experiences, the obtained verbal reports were classified into three broad categories. One category is the *recall of a distinct experience*: an experience that relates to a particular event with beginning and end. According reports were indicative of the constructive mode: recalling contextual information from a specific experience was followed by the reconstruction of the value judgment from the recalled facts. For example: *"The reason I got this device was to develop applications for it. [the company] has a special program for educational institutions to which provides free licenses for development. But when we contacted them, they even questioned the existence of our institution... this should have happened around here [points to the curve]"*. Such distinct reports provided one or more contextual cues about the past experience, such as temporal information (i.e. when the event took place), changes in the context of use (e.g. "then I went on vacation..."), information related to the participant's social environment (e.g. "a friend of mine was very positive..."), etc. They constituted the most dominant type of reporting (37 of 74, 50%).

Other reports provided no contextual information about a recalled experience, but instead, the participant reported an overall evaluation without further motivation: *"after that, [my opinion] is about constant, it hasn't changed lately"*. Such reports are typical for a pure value-account mode of recall: recalling an overall evaluation of a specific experience or period, while failing to recall contextual cues or facts about an exact experience (24 of 74, 32%).

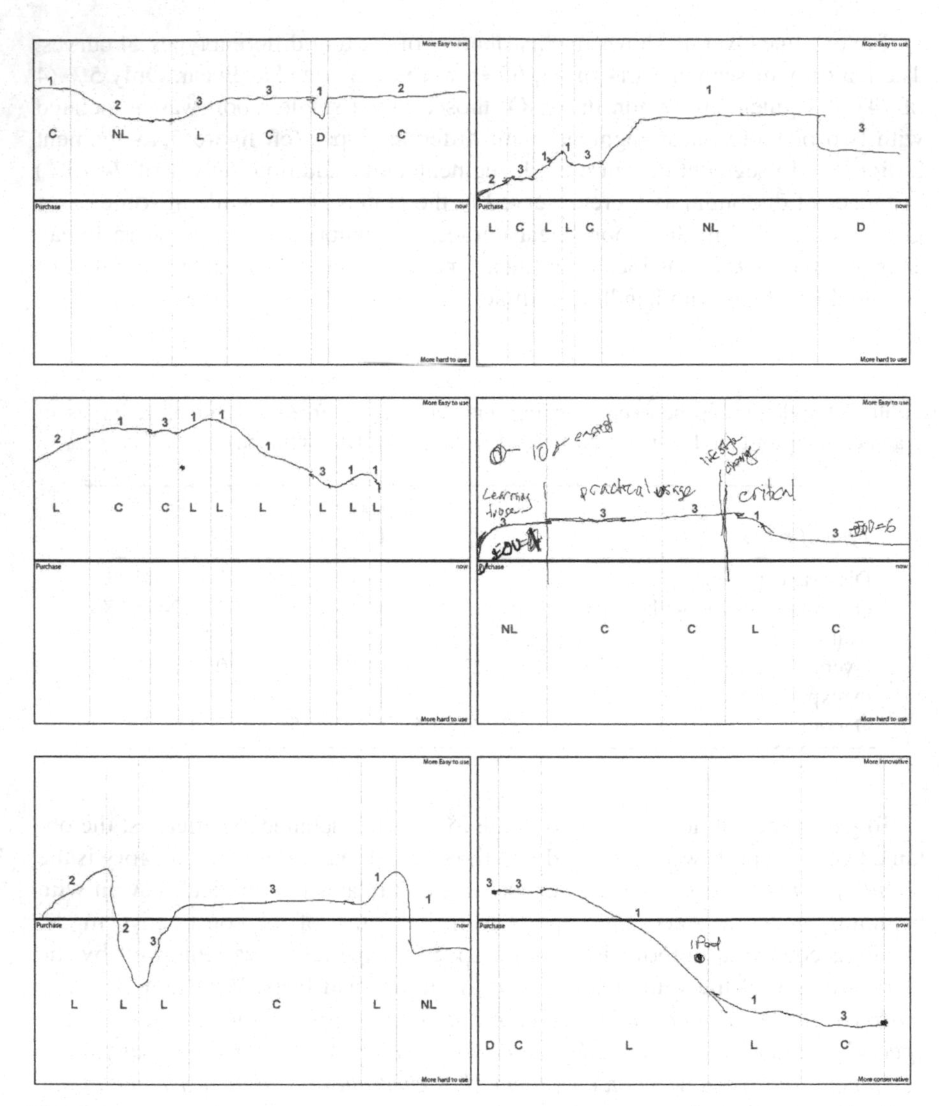

Fig. 5.4 Examples of free-hand graphing (left: a,c,e. right: b,d,f). Identified segments are indicated by vertical lines. Each segment is coded for the type of report (1: Reporting a discrete experience, 2: Reporting an overall evaluation, reasoning through experience, 3: Reporting an overall evaluation with no further reasoning) and type of graph (C: Constant, L: Linear, NL: Non-Linear, D: Discontinuous).

We further found a third type of reporting that combines the two types mentioned above: *"[my opinion] decreased as I expected that it would be easier than that, for example, I would like to have the automatic tilting to landscape view as it has an accelerometer"*. Those reports were grounded in the recall of an overall evaluation, but participants proceeded to reason about this value-judgment through reporting specific experiences (13 of 74, 18%). Most of them (10 of 13) reflected linear changes.

5.3.2.2 How Does iScale Compare to Free-Hand Graphing?

The two iScale tools were also compared to free-hand graphing. Participants' verbal reports were transcribed and analyzed using Conventional Qualitative Content Analysis (Hsieh and Shannon, 2005). We started with *open coding* where we aimed at identifying an overpopulated list of design properties that appear to influence the design space of the three graphing techniques. Subsequently, we grouped the initial codes into overall categories through an iterative process. Each statement was coded for the property the participant mentions as well as to whether or not this property affects the graphing or the recalling process. Statements were always differentiating two of the approaches from a third (e.g., the two iScale versions as opposed to FHG) as this was imposed by the structure of the employed Repertory Grid interview technique (e.g., "think of a property or quality that makes two of the graphing techniques alike and discriminates them from the third").

Table 5.3 illustrates the dominant properties that were elicited in the interview sessions. For each property, it displays the number of participants mentioning it as present for a given technique, and the number of participants mentioning it as affecting the graphing or recalling process. The design properties can be distinguished into three broad groups: *expressiveness*, *control*, and *Interplay graphing-recalling*.

5.3.2.3 Expressiveness

As expected, the majority of users perceived the free-hand graphing approach as more expressive. This was mostly due to the ***freedom in graphing*** that the free-hand approach provides as opposed to the iScale tools that restrict the user in plotting points connected through line segments.

The majority of participants emphasized the effect this has on the graphing activity. While all participants expressed the perceived benefit of FHG for graphically expressing their opinions, only one participant mentioned that this also affects the recalling process as the graph provides richer feedback and therefore more cues for recalling. One participant stated freedom in graphing as a positive property for expressing impressions for which she fails to recall any supportive facts.

Next, some participants mentioned the ability to ***annotate*** the graph as positive , because it enhances the recalling process. Annotations helped in recollecting *contextual* and *temporal* cues from the past, such as positioning a specific experience along the timeline, splitting the timeline into periods, but also in externalizing thoughts that participants thought they might fail to recall afterwards.

Table 5.3 Design properties elicited in the interview sessions. Number of participants using a given property to differentiate the three graphing techniques, and number of participants mentioning the given property as affecting the graphing or recalling process.

Name	Description	Tool			Impact on	
		FHS	CON	VA	Graph	Recall
Expressiveness						
Graphing freedom	Increasing the degrees-of-freedom in graphing	7			6	1
Annotation	Enabling free-hand annotation on the graph	3				3
Control						
Interactivity	Providing instant visual feedback through quick actions		8	8	4	4
Modifiability	Providing the ability to modify the graph as new experiences are being recalled		8	6	8	2
Temporal Overview	Providing feedback on overall opinion across full timeline			7	3	4
Interplay graphing-recalling						
Chronological linearity	Chronological linearity in recalling	5	5		1	5
Concurrency	Concurrency of time in graphing & recalling	5	5		2	5

5.3.2.4 Control

Most participants stated that the iScale tools provide more overall control. First, eight out of the twelve participants found the constrained interaction a positive aspect of iScale, providing better ***interactivity***, as it consumes less resources, thus providing them *better control of the output* (4 participants) and *enabling them to focus on recalling their experiences* (4 participants).

Second, participants differentiated iScale from FHG in terms of the ability to modify the graph while new experiences are recalled. Some participants further differentiated between the two iScale tools in terms of modifiability.

Third, seven out of the twelve participants acknowledged that the value-account tool provides a better overview of the full timeline, and, thus, ***temporal overview*** enhancing their recall process.

5.3.2.5 Interplay Graphing-Recalling

Five participants in total mentioned ***temporal linearity*** as a property that differentiated free-hand graphing and the constructive iScale from the value-account tool. Most of those participants mentioned that recalling events in a step-by-step order helped them in recalling more events, while some of them were negative towards value-account as they felt that it constrained them when recalling events due to a focus in compiling a coherent story.

Similarly, five participants highlighted that the ***concurrency*** of graphing and reporting, that was a feature of the constructive version of iScale, enhances the output of both the recall and the graphing process.

5.3.3 Discussion

As expected, free-hand graphing was found to be more expressive than iScale due to the increased degrees of freedom in graphing as well as due to its ability to easily annotate graphs. However, participants also reported properties that were not present in the FHG, such as the two-step interactivity and modifiability of the electronic graphs that resulted in a better interoperability between the graphing and the recalling activity. Participants also reported benefits for both the constructive and the value-account iScale. The value-account version provided a temporal overview, which influenced both graphing and recall. The constructive approach provided benefits, such as the chronological order and concurrency between graphing and reporting which had a positive impact on the recall process.

The need for graphing non-linear and discontinuous curves was limited, while most non-linear curves could be approximated by two linear segments. The need for annotation was highlighted by participants in the post-use interviews and two forms of annotation were added to the tool: (a) a timeline annotation that allowed users to set the start and end date of graphed segments, thereby splitting the timeline in specified periods, and (b) a visualization of experiences along the respective line segment that they belong to, with a brief identifier for the experience (see figure

5.2a). Annotation provided users with the ability to combine an overview of the different periods as well as the experiences that defined these periods. Annotation also promotes interactivity as users have a better overview of the graphed pattern and are therefore more likely to modify it.

5.4 Study 2: Benefits and Drawbacks of the Constructive and the Value-Account Version of iScale

While iScale appeared to be a viable alternative to free hand graphing, the comparative benefits and drawbacks of both iScale variants merited a second study. We compared the constructive and the value-account version of iScale to a control condition that entailed reporting one's experiences with a product without any support through graphing. We focused on the number, the richness, and the test-retest consistency of the elicited experience reports.

5.4.1 Method

5.4.1.1 Participants

Forty-eight individuals (17 Female, Median Age=23, Min=18, Max=28) participated in the experiment. They were all students at a technical university and were rewarded for participating in the experiment; 19 of them majored in management related disciplines, 16 in design, and 13 in natural sciences and engineering. They all owned a mobile phone for no less than four and no more than eighteen months; 16 participants owned a smart phone. No significant differences were found between participants in the constructive and the value-account condition in length of ownership (M_{con} = 13 months, M_{va} = 10 months, t(46)=1.51, p=.13) and type of mobile phone (5 participants owned a smart phone in the constructive condition, 11 participants in the value-account, χ^2=2.3, p=.13).

5.4.1.2 Materials

Three different versions of iScale were used in the experiment: *constructive*, *value-account*, and *no-graphing* (control). The constructive and value-account versions employed the two distinct graphing approaches described earlier. No-graphing was a stripped-down version of iScale, with the graphing interface completely removed. Thus, users were only provided with the interface to report experiences (see figure 5.3a) and this was used as a control condition to test the effect, if any, of graphing.

5.4.1.3 Study Design

A 3x2 study design was employed with *mode of recall* (constructive, value-account, no graphing/control), and *product quality* being reported (i.e. ease-of-use versus innovativeness) as independent factors.

Mode of recall

Product quality	Constructive	Value-Account	Control
Ease of use	A	B	C
Innovativeness	D	E	F

Fig. 5.5 Study design

5.4.1.4 Procedure

Participants joined two sessions, each one lasting approximately 40 minutes and separated by approximately one week (minimum: 7 days, maximum: 10 days). During the first session, participants used two different tools (either of the two graphing versions of iScale and the no-graphing version) to report on two qualities of their mobile phones (see section 5.3.1.2 for a motivation of the chosen product qualities). For instance, participants in condition 1 (see figure 5.6) used the constructive iScale to report on ease of use, followed by the no-graphing tool to report on innovativeness. During the second session, participants used the same combinations of tool - product quality, but in reverse order (see figure 5.6). As such, participants' consistency across those two sessions could be used a measure of test-retest consistency of the recall process.

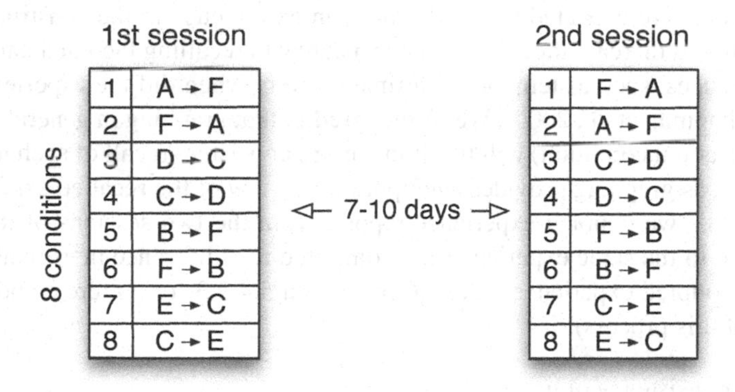

Fig. 5.6 Study procedure. Participants joined two sessions. In each session they used two different tools to report on two different product qualities. Tasks A to F refer to the six conditions of the study design (see figure 5.5).

5.4.1.5 Dependent Variables and Expectations

Despite the fact that the study was explorative in nature, a number of predictions about the differences in performance of the three versions of the tool can be made.

Number of elicited experience reports

Based on existing evidence that the reconstruction of events in a serial chronological order cues the recall of temporally surrounding experiences and related contextual cues (Anderson and Conway, 1993), it was expected that the constructive iScale will result in an increase in the number of experiences being reported. For the value-account iScale, which makes it more difficult for participants to reconstruct their experiences in a chronological order, the difference to the control condition was expected to be smaller.

Richness of elicited experience reports

Similar to number of elicited experience reports, we expected that reconstructing in a chronological order would lead to more contextual cues, thus providing richer insight into users experiences. Such contextual information may relate to temporal (i.e. when did the event happen), factual (i.e. what happened), social (i.e. who was present) and others. To identify these different factors of richness, we submitted the experience reports to a qualitative content analysis (Hsieh and Shannon, 2005) (see section 5.4.2.2 for a more elaborate description of this process).

Test-retest consistency in time estimation

As participants are expected to recall more contextual cues in the constructive iScale, this should increase the test-retest consistency in recalling factual details of the past experiences, such as temporal information (e.g. when did the experience take place) (Kahneman et al., 2004). We further predict that graphing in general (even in the value-account condition) will result in a more consistent recall of such temporal information, as graphing provides a temporal overview of the recalled experiences. To assess this, we coupled experience reports from the two sessions of the study that referred to the same experience and computed the difference in estimated time across the coupled experience reports (see section 5.4.2.3 for a more elaborate description of this process).

Test-retest consistency of graphs

The test-retest consistency of the participants' graphs (i.e. value-charged information) was expected to be higher in the value-account version, where participants cue this directly through a hypothetical memory structure, compared to the constructive version, where participants are assumed to reconstruct this information from concrete contextual cues recalled from episodic memory. This is based on the assumption that repeated chronological reconstruction might cue a different set of experiences and, thus, lead to a different path in reconstructing the overall pattern of how one's experience with a product developed over time. As a measure of the test-retest consistency of the graphs, we computed the area between the two graphs through sampling them in 100 steps (see figure 5.9).

5.4.2 Analysis and Results

A total of 457 experience reports were elicited. Participants provided an average of 4 to 6 experience reports depending on the recall condition. Ninety-five percent of all experiences were related to the first six months of use. We compare the two graphing tools to the no-graphing (control) condition in terms of a) the number of elicited experience reports, b) the richness of elicited experience reports, c) the test-retest consistency in time estimation, and d) the test-retest consistency of graphs.

5.4.2.1 Number of Elicited Experience Reports

Figure 5.7 shows the number of reported experiences as a function of the mode of recall. An average of 6.1 experience reports was elicited when using the constructive iScale, 4.6 when using the value-account iScale, and 4 when using the no-graphing (control) tool.

An analysis of variance with number of experience reports as dependent variable and mode of recall (constructive, value-account, control), and product quality (ease-of-use, innovativeness) as independent variables revealed a significant main

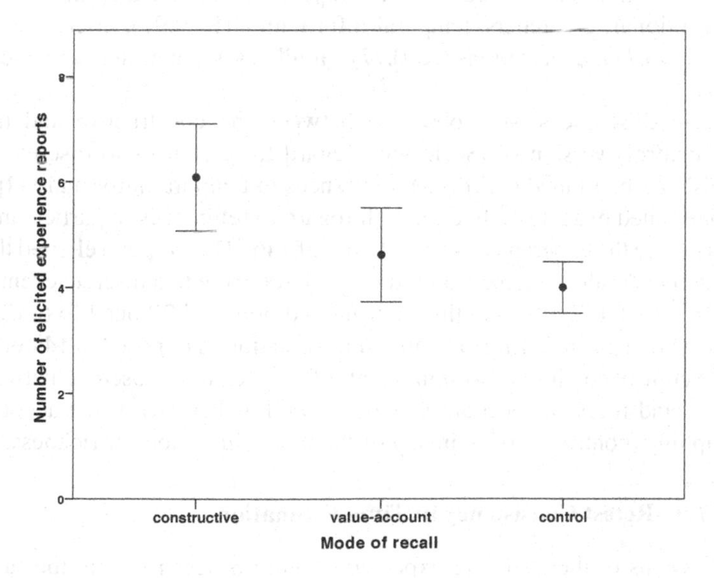

Fig. 5.7 Average number of experience reports, together with their 95% confidence intervals, elicited by participants using the constructive, the value-account, and the no-graphing (control) version of iScale.

effect for mode of recall, $F(2,89) = 7.74$, $p<.05$, $h_p^2=0.15$, but not for product quality, $F(1,89) = 1.64$, $p=.2$, $h_p^2=0.02$, or for the interaction between mode of recall and product quality, $F(2,89) = 1.19$, $p=.7$, $h_p^2=0.007$. Post-hoc tests using the Bonferroni correction revealed that participants in the constructive condition elicited a significantly higher number of experience reports than in the value-account (p=.04, Cohen's d=0.7) and the control condition (p<.001, d=1). No significant differences were demonstrated between the value-account and the control condition (p=0.67, d=0.3).

5.4.2.2 Richness of Elicited Experience Reports

To identify the different factors of richness, we submitted the experience reports to a qualitative content analysis (Hsieh and Shannon, 2005). Open coding was performed by the first author and resulted in three main types of information present in the reports: *discrete event information* summarizes references to a concrete occurrence that influenced the experience (e.g., "I found out that there was a mail application"), *temporal information* which summarizes references to the exact time at which the experience took place (e.g., "The first week I tried to practice the touch screen"), and, *expectations* which summarize references to participants' expectations about the reported experience (e.g., "As I bought this phone in Europe I expected that at least all European languages are available for free online"). Each report was coded by the first author for the presence or absence of temporal information, discrete event information and expectations. Interrater agreement (Fleiss' Kappa, 2003) was computed on a small random sub-set of the reports (10%) coded by the first author and two additional researchers: temporal information (K=0.97), discrete event information (K=0.71), expectations (K=0.77). In all cases, interrater agreement was satisfactory.

Significant differences were observed between the constructive and the no-graphing (control) version of iScale with regard to references to discrete events (p<.05, Fisher's one-tailed exact test), references to temporal information (p<.05, Fisher's one-tailed exact test), but not with regard to references to participants' expectations before the experienced event. 45 out of 146 (31%) reports elicited through the constructive iScale contained at least one cue referring to a discrete event as opposed to 38 out of 192 (20%) in the control condition, and 20 out 146 (14%) contained at least one cue referring to temporal information as opposed to 14 out of 192 (7%) in the control condition. No significant differences were observed between the constructive and the value-account version as well as between value-account and the no-graphing (control) version in any of the three dimensions or richness.

5.4.2.3 Test-Retest Consistency in Time Estimation

The two sessions of the study are expected to more or less result in the same experience reports. Thus, reports across these two sessions can be coupled. A total of 325 experience reports (71%) were coupled. We only coupled reports when we had sufficient confidence that they reported the same experience. This, inevitably, left

Table 5.4 Number of experience reports judged for the three dimensions of richness, elicited through the three different versions of iScale: constructive, value-account and no-graphing (control).

Name		Constr.	VA	Control
Contextual information				
a. Event: Does the participant recall one or more discrete events that lead to the realization of the reported experience?	Y:	45	27	38
	N:	101	91	154
b. Temporal: Does the participant recall temporal information about the reported experience?	Y:	20	16	14
	N:	126	102	178
c. Expectation: Does the participant recall his/her expectations about the reported experience?	Y:	10	11	16
	N:	136	107	176

out short reports that did not contain sufficient text to form this judgment. 72% of reports in the constructive condition were coupled, 81% in the value-account and 69% in the control condition. 52 coupled reports had incomplete time information (in either session), leaving a total of 273 complete coupled reports.

For each reported experience participants estimated the moment in time (i.e. days, weeks or months after the purchase of the product) at which the experience took place (see figure 5.3a). We used formula 5.1 to compute the distance (or convergence) between the recalled points in time of both sessions (for a justification of the logarithmic transformation see the appendix section).

$$\Delta = Abs(log(t_2) - log(t_1)) \tag{5.1}$$

An analysis of variance with the computed temporal distance Δ between experience reports from session 1 and session 2 as dependent variable and mode of recall (constructive, value-account, control) and product quality (ease-of-use, innovativeness) as independent variables displayed significant main effects for mode of recall, $F(2, 267) = 5.42$, $p<.01$, $h_p^2=0.04$, and for product quality, $F(1,267) = 4.66$, $p<.05$, $h_p^2=0.02$, but not for the interaction between mode of recall and product quality, $F(2,267) = 1.81$, $p=.2$, $h_p^2=0.01$. Post-hoc tests using the Bonferroni correction revealed that participants in the constructive condition were significantly more consistent in estimating the time that an experience took place as compared to the control condition ($p<.01$, Cohen's $d=4.41$). No significant differences were established between the two graphing conditions ($p=.91$, $d=1.29$), or between the value-account and the control condition ($p=0.1$, $d=3.01$). Last, a significant difference was found in the temporal consistency of reports for both product qualities, where reports of ease-of-use were temporally more consistent than reports of innovativeness, $t(271)=2.8$, $p<0.01$, Cohen's $d=3$.

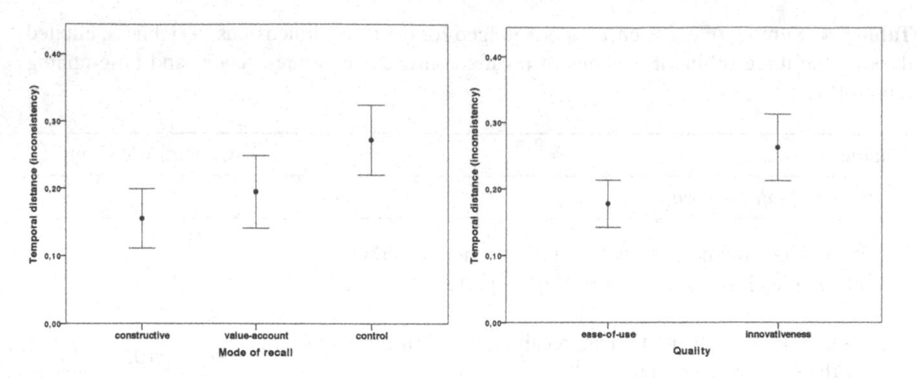

Fig. 5.8 Mean temporal inconsistency, together with the 95% confidence interval, between two recalls of the same experience, (left) when using the constructive, the value-account, and the no-graphing (control) version of iScale, (right) when reporting on perceived ease-of-use and perceived innovativeness.

5.4.2.4 Test-Retest Consistency of Graphs

Figure 5.9 displays example graphs by two participants in two respective sessions with the constructive iScale and two participants with the value-account iScale. The area between the two graphs is a simple measure for the inconsistency in participants' graphs. It was calculated through sampling the graphs in 100 steps.

An analysis of variance with this area measure as dependent variable and mode of recall (constructive, value-account) and product quality (ease-of-use, innovativeness) being reported as independent variables revealed a significant effect for mode of recall, $F(1, 44) = 6.75$, $p<.05$, $h_p^2=0.13$, but not for product quality, $F(1, 44) = .481$, $p=.49$, $h_p^2=0.01$, or for the interaction between mode of recall and product quality, $F(1,44) = 0.073$, $p=.8$, $h_p^2=0.002$. Graphs elicited through the constructive iScale tool were more consistent ($M_{area}=30.1$, $SD=17.4$) than ones elicited through the value-account iScale ($M_{area}=52.2$, $SD=36.7$, $t(46)=2.7$, $p<.05$, cohen's $d=0.8$)

5.4.3 Discussion

As expected, the constructive iScale tool lead to a significant increase in the number of elicited reports (with a large effect size, d=1, Cohen, 1992), but also the richness, of elicited reports, as compared to the control condition that involved no graphing. When using the constructive iScale, participants elicited approximately 50% more experience reports, and were more likely to recall cues referring to a discrete event, or ones referring to temporal information, such as when the experience took place. Contrary to our expectations, the value-account iScale did not result in similar benefits. These findings support the idea that graphing, through imposing a chronological order in the recall process (Anderson and Conway, 1993), supports the reconstruction of the context in which experiences took place, thus forming stronger temporal and semantic links across the distinct experiences (Kahneman et al., 2004).

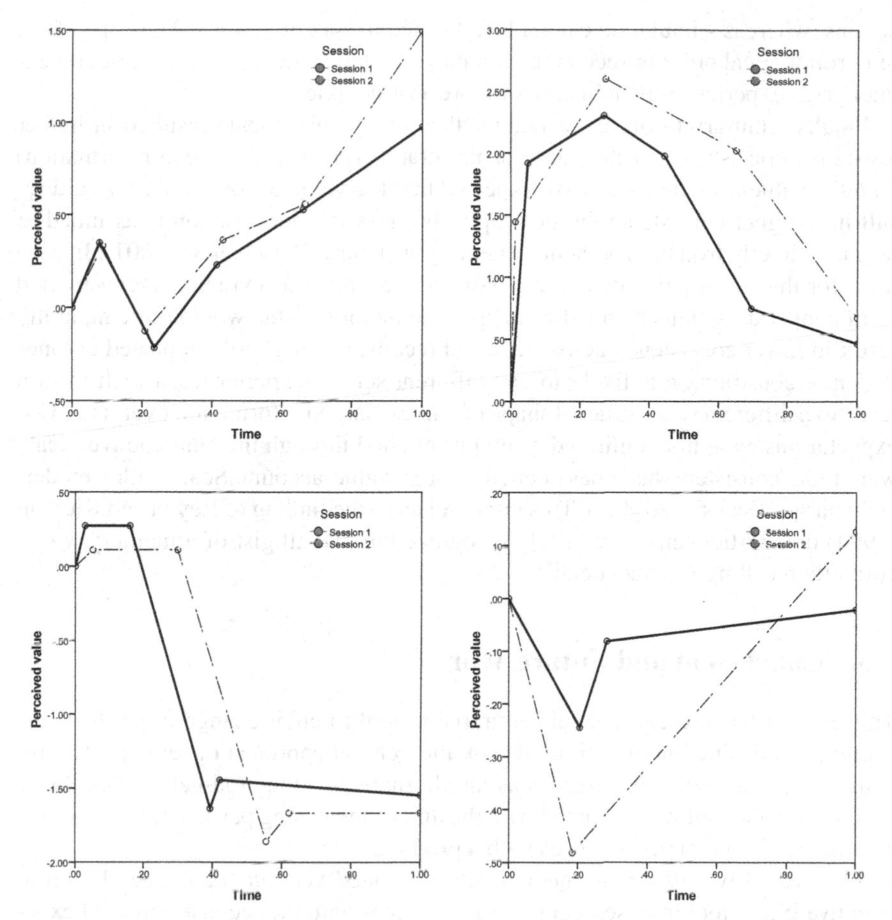

Fig. 5.9 Example graphs elicited in the constructive (top) and the value-account (bottom) conditions during the two sessions.

Next to an number and the richness of elicited reports, the constructive iScale tool also demonstrates the highest consistency across the two sessions (with a large effect size, d=4.41, Cohen, 1992), in recalling the exact time that each experience took place. These findings seem to be in line with the previous ones, i.e., given that the constructive iScale lead to a more effective reconstruction of the context in which experiences took place, this should also be beneficial when estimating when these experiences took place. Moreover, we found experiences relating to the ease-of-use of the product to be more reliably recalled (in terms of time estimation) in comparison to ones relating to the products' innovativeness. One possible interpretation might tap into the different nature of experiences that relate to ease-of-use and innovativeness. Von Wilamowitz-Moellendorff et al. (2006; 2007) similarly observed that participants often recall with greater ease contextual cues about experiences relating to ease-of-use rather than stimulation. Ease of use is tied to concrete actions, whereas stimulation cannot be allocated to specific events. Thus, the effect of chronological order in reconstruction may be more salient in the case of contextually rich experiences than in case of more abstract ones.

Finally, contrary to our expectations, the constructive iScale resulted in higher test-retest consistency of the participants' graphs (i.e. value-charged information) than the value-account iScale. We expected that the value-account iScale would result in a higher consistency in the graphs, because this information is assumed to be cued directly from a hypothetical memory structure (Betsch et al., 2001). In contrast, for the constructive iScale, we assumed that this information is reconstructed from contextual details recalled from episodic memory. This would quite naturally result in lower consistency across repeated recalls, as long as the repeated chronological reconstruction is likely to cue different sets of experiences, which in turn leads to a different reconstructed shape of value-charged information over time. Our expectations were not confirmed as graphs elicited through the constructive iScale were more consistent than ones elicited through value-account iScale with a moderate to large effect size (d=0.8). This might relate to the finding of Reyna and Kiernan (1994) that participants may falsely recognize the overall gist of a memory, while correctly recalling its exact details.

5.5 Conclusion and Future Work

This chapter has presented iScale, a graphing tool to elicit change in product perception and evaluation over time. It took the general approach of retrospective reconstruction of users' experiences as an alternative to longitudinal studies. More specifically, the tool was designed with the aim of increasing participants' effectiveness in recalling their experiences with a product.

We created two different, theoretically grounded versions of iScale. The constructive iScale tool imposes a chronological order onto the reconstruction of experiences. This will increase the contextual cues surrounding the experienced events, cues used to reconstruct value-charged information (i.e., affect). The value-account iScale aims at distinguishing the elicitation of the two types of information: the value-charged and contextual cues. It assumes that value-charged information can be recalled from a separate, specific memory structure.

Study 1 provided qualitative insights into the use of iScale and compared it to free-hand graphing. The study highlighted the importance of annotations and led to a redesign of iScale with two new forms of annotation, (a) timeline annotation that allowed users to set the start and end date of graphed segments, and (b) a visualization of experiences on the main, graphing, interface that allowed an overview of one's storyline. Overall, the study confirmed our expectations that free-hand graphing is more expressive than iScale due to the increased degrees of freedom in graphing as well as due to its ability to easily annotate graphs. Nevertheless, participants reported qualities present in iScale, such as two-step interactivity and modifiability of the electronic graphs, which resulted in a better interoperability of graphing and the recalling activity.

Study 2 tested the effectiveness of graphing with the two different versions of iScale against a control condition, which allowed for the direct reporting of experiences, without any form of graphical representation. Overall, we found that the constructive iScale provided better assistance than the the value-account iScale tool in the reconstruction process. When using the constructive iScale, participants elicited approximately 50% more experience reports than in the control (no-graphing) condition, were more likely to recall cues referring to a discrete event, or ones referring to temporal information such as when the experience took place, and were more consistent across multiple recalls in estimating the time when each experience took place. Moreover, contrary to our expectations, participants graphed patterns were more consistent in the constructive than in the value-account condition. These findings support the idea that graphing, through imposing a chronological order in the recall process (Anderson and Conway, 1993), supports the reconstruction of the context in which experiences took place, thus forming stronger temporal and semantic links across the distinct experiences (Kahneman et al., 2004).

While iScale is promising, one crucial aspect must be kept in mind. There is a likely discrepancy between experiences elicited through longitudinal field studies composed of records of moments (e.g., through experience sampling, Hektner et al., 2007) and retrospective data elicited from memory through iScale. Retrospective reconstructions cover long periods of time and, thus, systematic biases, such as the overemphasis on especially salient moments, are likely to occur (Kahneman, 1999; Bartlett, 1932). The current studies provide no insights into how these retrospections differ from the actual experiences and future work should inquire in those differences with actual longitudinal studies.

In this chapter we argued that veridicality may not be as important as the test-retest consistency of the recall process, because people actually communicate and act upon their own biased memory and not on an unbiased objective summary of what actually happened. In supporting design, understanding what users remember may be more important than what they actually experienced. Still, designers are not always interested in users memories. Often, the actual, and not the remembered, experiences should be at the forefront. Consider, for instance, the case where we might want to know the reasons that underly non-responsible driving behavior. Memories offer little understanding as to what motivates such behaviors. Retrospective techniques are not aimed to replace longitudinal field studies and in-situ methods. Instead, we propose that retrospective techniques may be a viable alternative to longitudinal studies when memories are placed at higher importance than actuality.

Next, in our study, we assessed users' test-retest consistency in recall using two different pieces of information: the exact time that an experience took place, and the overall graphed pattern. These two served to indicate the two different kinds of recalled information: episodic and value-charged (i.e., affect). The former, we thought, would signify the effective recall of a substantial amount of contextual cues from episodic memory. We expected that estimating this temporal information would be an error-prone activity. Thus, a more effective recall would have a strong impact on users' consistency in time estimation, through the presence of more contextual cues.

The latter, we thought, would signify the recall of experiential information (i.e., affect), either through inferring this from recalled episodic information (Robinson and Clore, 2002), or through recalling this directly from value-account Betsch et al. (2001). These two metrics of consistency are not without limitations; future work should expand to different facets of consistency.

Beyond the cost-effective elicitation of longitudinal data, this work provides support for the viability of survey methods that guide participants through a structured process of data elicitation. A wealth of such procedures exists for face-to-face interviewing. For instance, structured interview techniques such as triading (Fransella et al., 2003) and laddering (Reynolds and Gutman, 1988) imprint a particular structure onto the data elicited by participants, which makes the data computational friendly. Another example is Vermeersch' explicitation technique (see Light, 2006), which employs a particular style of interviewing that aims at supporting the interviewee to enter a state of evocation and "relive" the activity under consideration. While such techniques, when used in face to face interviewing, are most powerful, they are labor-intensive and require skilled interviewers (Groves et al., 2009), which always constrains the sample size of the study. Self-reporting approaches, on the other hand, can have impact because one can survey large samples and, by that, also inquire into rare experiences and atypical behaviors.

Obviously, iScale can as well produce large amount of qualitative information that will require labor-intensive analysis given traditional qualitative data analysis procedures like Content Analysis (Krippendorff, 2004; Hsieh and Shannon, 2005) and Grounded Theory (Strauss and Corbin, 1998). Novel techniques from the field of information retrieval (Landauer and Dumais, 1997; Blei et al., 2003) may prove especially fruitful in automating or semi-automating the qualitative analysis process. Finally, the interpersonal analysis of the graphs is definitely a subject for further research and was addressed here only superficially.

iScale was motivated by a need for lightweight methods that provide insights into long-term usage and related experiences. While the importance of temporality has been repeatedly highlighted in user experience research (Forlizzi and Battarbee, 2004; Hassenzahl and Tractinsky, 2006), it has rarely been systematically addressed. In two recent studies (Karapanos et al., 2008a, 2009c) we provided some first evidence that not only our perceptions, but also the relative weight of different product qualities change over time. So far, both academia and industry have largely focused on initial use. This has strong implications for the quality of interactive products. For instance, Den Ouden et al. (2006) found that an alarmingly increasing number of returned products, in 2002 covering 48% of all returned products, are technically fully functional (i.e. according to specifications), but they are returned on the basis of failing to satisfy users' true needs (28%), or purely on users' remorse (20%). These failures were not so much related to problems rooted in early interactions - problems that can often be overcome through learning -, but to those that persist over time, pointing at failures to truly incorporate the product into daily life. We hope that iScale provides a first step towards retrospective elicitation methods as a viable, lightweight alternative to the expensive longitudinal methods. Only by that, we can start to fully account for the notion of User Experience as a temporal phenomenon.

5.6 Appendix - Temporal Transformation

In section 5.4.2.3 we used users' consistency of temporal information of reported experiences across repeated recalls as a metric of reliability of the recall process. One question, however, relates to whether participants' accuracy in recalling temporal information remains constant across the full timeline, from the moment of purchase of the product to the present time. The participant's accuracy might be affected by the amount of contextual information surrounding the experience that is available at the moment of recalling. Theories of recall have suggested that recent experiences (Koriat et al., 2000), or experiences associated with important milestones (e.g. the purchase of the product) (Barsalou, 1988) might be more easily accessible. If such biases exist, they will affect the reliability test as differences in the consistency of recalled information might be due to pertaining to more or less salient periods and not due to the reconstruction process. In the presence of such biases, the temporal distance between the two coupled experience reports elicited in the two distinct sessions should be transformed to account for the accessibility biases.

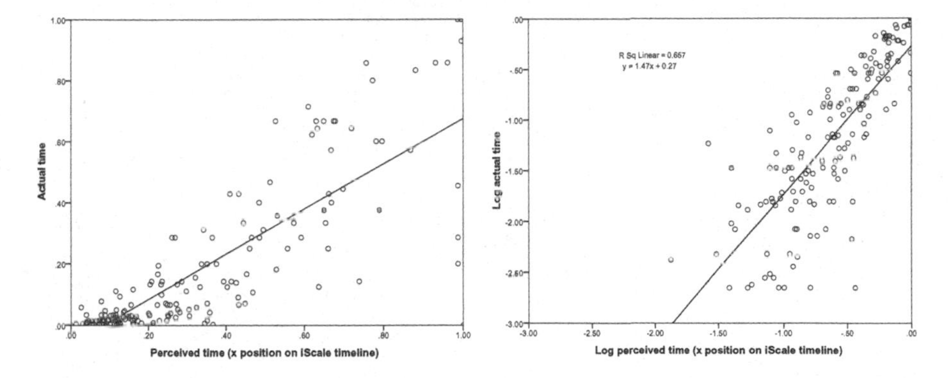

Fig. 5.10 Relation between actual time (reported time for a sketched node) and position of node along the iScale's x-axis: (a) linear relation, (b) power-law relation. Actual time (days) is adjusted to the full time of ownership of the product for each participant.

We attempt to assess the existence of accessibility biases through examining the way in which participants used the timescale of the tool, (i.e. iScale's x-axis). Participants graphed linear curves through adding nodes in the graph (see figure 5.2a). Each node can be characterized by two properties: a) the actual time (participants explicitly annotated for each node the approximate amount of days, weeks, or months after purchase that this node represents, and b) the perceived time (the position of the node along the x-axis of iScale).

Figure 5.10 depicts the relationship between the reported (actual) time versus the perceived time (i.e. the position of the node along iScale's x-axis). To enable an across-subject comparison, we normalized the reported (actual) time variable by the total time of ownership of the product for each participant, resulting to an index from 0 to 1. Given no accessibility bias, one would expect a linear relationship between these two pieces of information. One might note in figure 5.10a, however, that the variance in the dependent variable (actual time) is not uniformly distributed across the range of the independent variable (position along the x-axis of iScale). If one transforms the variables by the logarithmic function, the data become much more uniformly distributed. A linear regression on these transformed variables shows a significant prediction accounting for 66% of the variance in the dependent variable. This suggests a power law relation between the recalled actual time of the experienced event and its position along the graphing tool's timeline with a power equal to 1/1.47=0.68 (i.e. perceived-time = actual-time$^{0.68}$). In other words, participants had a tendency to use a substantial fraction of the x-axis of iScale to map their initial experiences. In a similar vein, 95% of all experience reports related to the first six months of use and 75% of all experience reports related to the first month of use whereas the median time of ownership was 10 months. It thus becomes evident that experiences pertaining to initial use are more accessible in participants' memory. To account for this accessibility bias we compute the temporal distance between two events through formula 5.1.

Chapter 6
A Semi-Automated Approach to the Content Analysis of Experience Narratives

Abstract. iScale will typically result in a wealth of experience narratives relating to different stages of products' adoption. The qualitative analysis of these narrative is a labor intensive, and prone to researcher bias activity. This chapter proposes a semi-automated technique that aims at supporting the researcher in the content analysis of experience narratives. The technique combines traditional qualitative coding procedures (Strauss and Corbin, 1998) with computational approaches for assessing the semantic similarity between documents (Salton et al., 1975). This results in an iterative process of qualitative coding and visualization of insights which enables to move quickly between high-level generalized knowledge and concrete and idiosyncratic insights. The proposed approach was compared against a traditional vector-space approach for assessing the semantic similarity between documents, the Latent-Semantic Analysis (LSA), using a dataset of a study in chapter 4. Overall, the proposed approach was shown to perform substantially better than traditional LSA. However, interestingly enough, this was mainly rooted in the explicit modeling of relations between concepts and individual terms, and not in the restriction of the list of terms to the ones that concern particular phenomena of interest.

6.1 Introduction

The previous chapter proposed a survey technique for the elicitation of experience narratives from a large sample of users. The researcher is then faced with an overwhelming amount of idiosyncratic experience narratives. Each narrative may provide a rich insight into the experience and the context in which it takes place. However, generalized knowledge may also be gained from these experience narratives. Such generalized knowledge may be reflected in questions like: how frequent is a certain kind of experience, what is the ratio of positive versus negative experiences and how does this compare to competitive products, how does the dominance of different product qualities fluctuate over time and what should we improve to motivate prolonged use?

E. Karapanos: Modeling Users' Experiences with Interact. Syst., SCI 436, pp. 115–136.
springerlink.com

Contrary to idiosyncratic insights, generalized knowledge requires a pre-processing step: that of assessing the similarity between different experiences. While it is well-acknowledged that every experience is unique and non-repeatable (Forlizzi and Battarbee, 2004; Hassenzahl and Tractinsky, 2006; Wright and McCarthy, 2008), different experiences share similar features: some might refer to the same product feature; others might be motivated by the same human need despite the seemingly different nature (Hassenzahl, 2008); others might refer to the same product quality; some are filled with positive emotions while others are dominated by negative ones. These features may form the basis for assessing the similarity of different experience narratives.

In assessing the similarity between different narratives, content analysis techniques (Krippendorff, 2004; Hsieh and Shannon, 2005) may be employed in identifying key concepts in the data, identifying a hierarchical structure among the concepts and classifying narratives into broad categories. Such approaches are laborious as the researcher needs to process all narratives in identifying the key concepts in the data.

A number of automated approaches to semantic similarity assessment have been proposed in the field of Information Retrieval and can potentially assist the analyst in this task. Such approaches typically rely on vector space models (Salton et al., 1975) in which the degree of semantic similarity between documents is related to the degree of term co-occurrence across the documents. As we will argue in this chapter, these approaches exhibit a number of limitations when one is concerned about analyzing self-reported experience narratives. First, they assume a homogeneity in the style of writing across documents which does not hold in this context as the vocabulary and verbosity of documents might substantially vary across different participants. Second, similarity is computed based on the co-occurrence of all terms that appear in a pool of documents while in the qualitative analysis of experience narratives the researcher is typically interested only in a limited set of words that refer to the phenomena of interest. As a result, words that are of minimal interest to the researcher may shadow the semantic relations that researchers are pursuing at identifying. Third, these automated approaches lack an essential part of qualitative research, that of interpretation. As different participants may use different terms or even phrases to refer to the same latent concept, an objectivist approach that relies purely on semantics will evidently fail in capturing the relevant concepts.

In this chapter, we propose a partially automated approach which combines traditional content analysis techniques (Strauss and Corbin, 1998) with computational approaches to assess the semantic similarity between documents (Salton et al., 1975). In identifying the concepts which will form the basis for computing the similarity between narratives, the proposed approach combines existing domain-specific knowledge with open-coding procedures that aim at identifying constructs that are not captured by existing measurement models. Domain-specific knowledge, on the other hand, for instance within the field of user experience, may be found in psychometric scales measuring perceived product qualities and emotional responses to products. Thus, only concepts that are relevant to the given context are considered in computing the similarity between narratives. At the same time, the process is automated as concepts identified in a small set of narratives may be used to assess the similarity among the full set of narratives. This results in an iterative process of coding and visualization of obtained insights.

Next, we describe the application of a fully-automated procedure that relies on a vector-space model (Salton et al., 1975), the Latent Semantic Analysis (LSA) and motivate the proposed adaptations of this approach towards a semi-automated approach.

6.2 Automated Approaches to Semantic Classification

A number of automated approaches exist for the assessment of semantic similarity between documents (for an extensive review see Kaur and Hornof, 2005; Cohen and Widdows, 2009). These approaches rely on the principle that the semantic similarity between two documents relates to the degree of term co-occurrence in these documents (Deerwester et al., 1990). In this sense, every document may be characterized as an n-dimensional vector where each element of the vector depicts the number of times that a given term appears in the document. The similarity between documents may then be computed in a high-dimensional geometrical space defined by these vectors.

Latent-Semantic Analysis (LSA) (Deerwester et al., 1990), also known as Latent-Semantic Indexing within the field of Information Retrieval, is one of the most popular vector-space approaches to semantic similarity measurement. It has been shown to reflect human semantic similarity judgments quite accurately (Landauer and Dumais, 1997) and has been successfully applied in a number of contexts such as that of identifying navigation problems in web sites (Katsanos et al., 2008) and structuring and identifying trends in academic communities (Larsen et al., 2008a).

LSA starts by indexing all n terms that appear in a pool of m documents, and constructs a $n \times m$ matrix A where each element $a_{i,j}$ depicts the number of times that the term i appears in document j. As matrix A is high-dimensional and sparse, LSA employs Singular-Value Decomposition (SVD) in reducing the dimensionality of the matrix and thus identifying the principal latent dimensions in the data. Semantic similarity can then be computed on this reduced dimensionality space which depicts a latent semantic space. Below, we describe in detail the procedure as applied in this chapter.

6.2.1 The Latent-Semantic Analysis Procedure

6.2.1.1 Term Indexing

Term-indexing techniques may vary from simple "bag-of-words" approaches that discard the syntactic structure of the document and only index the full list of words that appear in a document, to natural language algorithms that identify the part-of-speech, e.g. the probability that a term is a noun or a verb, in inferring the essence of a word (Berry et al., 1999). LSA typically discards syntactic information and treats each document as a pool of terms. However, it applies two pre-processing procedures in order to enhance the quality of the indexing procedure.

Firstly, a number of words, called *stop-words*, such as prepositions, pronouns and conjunctions, are commonly found in documents and carry no semantic information for the comprehension of the document theme (Fox, 1989). Such words are excluded from further analysis as they do not provide meaningful information and are likely to distort the similarity measure. We used a list stop-words provided by Fox (1989).

Secondly, the remaining terms are reduced to their root words through stemming algorithms. For instance, the terms "usability" and "usable" are reduced to the term "usabl", thus allowing the indexing of multiple forms of a word under one dimension in the vector-space model. We employed Porter's (1980) algorithm for stemming.

6.2.1.2 Normalizing Impact of Terms

The first step in the procedure has resulted in a $n \times m$ matrix A where each element $a_{i,j}$ depicts the number of times that the stemmed term i appears in document j. The frequencies of different terms across different documents will vary substantially. This results in undesired impacts of terms that are more frequent across a larger set of documents as they receive higher weight than terms that appear in only a small set of documents. However, these terms that appear in many documents have limited discriminatory power and are thus not very informative. One term-weighting criterion that counterbalances for this inherent bias is the term-frequency inverse-document frequency (TFIDF) (Salton and Buckley, 1988):

$$a_{i,j\_weighted} = a_{i,j} * log(nDocs/nDocs_i) \tag{6.1}$$

which weights the frequency $a_{i,j}$ by the logarithm of the ratio of the total number of documents $nDocs$ by the number of documents $nDocs_i$ in which the term i appears. Thus, frequent terms that appear in a large amount of documents and thus have little discriminatory power receive lower weight in the final matrix.

6.2.1.3 Dimensionality Reduction

Matrix A is sparse and high-dimensional. Moreover, certain groups of terms may display similar distributions across the different documents, thus underlying a single latent variable. LSA attempts to approximate A by a matrix of lower rank. Singular Value Decomposition is used to decompose matrix A in three matrices U, S, V in that

$$A = USV^T \tag{6.2}$$

Matrices U and V are orthonormal matrices and S is a diagonal matrix that contains the singular values of A. Singular values are ordered in decreasing size in matrix S, thus by taking the first $k \times k$ submatrix of S, we approximate A by its best-fit of rank k.

$$A_k = U_{nk} S_{kk} V_{mk}^T \tag{6.3}$$

6.2.1.4 Computing Document Similarity

The similarity between different documents or different terms may then be computed on the reduced dimensionality approximation of A. Matrices 6.4 and 6.5 constitute *m x m* and *n x n* covariances matrices for the documents and terms, respectively. The proximity matrices for the documents and terms are then derived by transforming 6.4 and 6.5 to correlation matrices.

$$S_R = A_k^T A_k = V_{mxk} S_{kxk}^2 V_{mxk}^T \tag{6.4}$$

$$A_k A_k^T = U_{nxk} S_{kxk}^2 V_{nxk}^T \tag{6.5}$$

Each element $s_{i,j}$ represents the similarity between documents, or terms *i* and *j*. The proximity matrix is normalized to a range (0,1) and transformed to a distance matrix with each element $d_{i,j} = 1 - |s_{i,j}|$.

6.2.2 Limitations of Latent-Semantic Analysis in the Context of Qualitative Content Analysis

Latent-Semantic Analysis has been shown to adequately approximate human judgments of semantic similarity in a number of contexts (Landauer et al., 2003; Katsanos et al., 2008; Larsen et al., 2008a). However, one may expect a number of drawbacks when compared to traditional content analysis procedures as applied by researchers.

First, LSA assumes a homogeneity in the style of writing across documents. Thus, the extend to which different words occur in one document over a second one denotes a difference in content across the two documents. This assumptions has been shown to hold in contexts of formal writing such as web pages (Katsanos et al., 2008) or abstracts of academic papers (Larsen et al., 2008a), but it is not expected to hold in qualitative research data such as interview transcripts or self-provided essays in diary studies as the vocabulary and verbosity of documents might substantially vary across different participants.

Second, LSA computes the similarity between documents based on the co-occurrence of all possible terms that may appear in the pool of documents. In the analysis of qualitative data, however, one is interested only in a small set of words that refer to a phenomenon that the researchers are interested in. As a result, words that are of minimal interest to the researchers may shadow the semantic relations that researchers are pursuing at identifying.

Third, LSA lacks an essential part of qualitative research, that of interpretation. As different participants may use different terms or even phrases to refer to the same latent concept, an objectivist approach that relies purely on semantics will evidently fail in capturing the relevant concepts. Ideally, automated vector-space models could be applied to meta-data that have resulted from open coding qualitative procedures (Strauss and Corbin, 1998). In the next section we propose such a semi-automated approach to semantic classification.

6.3 A Semi-automated Approach to Content Analysis

In this section we propose a semi-automated approach that aims at addressing the limitations of Latent-Semantic Analysis in assessing the similarity between self-reported experiences. The approach is different from Latent-Semantic Analysis in the following respects.

First, only terms relevant to the phenomena that the researcher is interested in are employed in the similarity measure. The approach exploits existing domain-specific knowledge in identifying relevant concepts, but also acknowledges that such knowledge will always be incomplete; thus the intervention of the researcher in extracting additional relevant concepts from the data is crucial.

Second, through manual coding the researcher is able to define explicit relations between concepts, i.e. latent dimensions, and individual measurement items, i.e. terms of phrases. For instance, an researcher may wish to code when the participant refers to a relevant person in a social context. The researcher may then create a concept named *'relevant others'* that is identified in the text through various individual terms such as *'friend'*, *'brother'*, *'colleague'* etc. Latent-Semantic Analysis would not be likely to identify these terms as semantically similar as these would not likely occur in the same documents. In contrast, through manual coding the researcher is able to explicitly relate these terms as different manifestations of the same latent concept.

Third, the approach utilizes visualization techniques in assisting the researcher in identifying the relevant concepts. For instance, an interacting visualization of the dissimilarity between narratives enables a systematic comparison of diverse narratives.

6.3.1 Incorporating Existing Domain-Specific Knowledge

The procedure starts by incorporating existing domain-specific knowledge. In characterizing users' experiences with interactive products, relevant knowledge might be found in psychometric scales measuring subjectively perceived product qualities (e.g. Hassenzahl, 2004; Lavie and Tractinsky, 2004; Hornbæk and Law, 2007) and emotional responses to products (e.g. Desmet and Hekkert, 2007).

Table 6.1 illustrates the five latent concepts that were used in the current study. The concepts were derived from (Hassenzahl, 2004) and (Lavie and Tractinsky, 2004). Each construct is measured through a number of individual (bi-polar) semantic differential scales (Osgood et al., 1957). Hassenzahl (2004) distinguishes *pragmatic* and *hedonic* quality in interactive products, while Lavie and Tractinsky (2004) differentiate between *classic* and *expressive aesthetics*. Both poles of all individual scales are used in defining each respective concept, after being stemmed to their root form (Porter, 1980).

Table 6.1 Domain-specific knowledge employed for the analysis of the 329 narratives. The domain-specific concepts were derived from Hassenzahl (2004) and Lavie and Tractinsky (2004) psychometric scales.

Concept	Individual terms
Pragmatic Quality	technical, human, complicated, simple, impractical, practical, cumbersome, direct, unpredictable, predictable, Confusing, clear, Unruly, manageable
HQ-Stimulation	typical, original, standard, creative, cautious, courageous, conservative, innovative, lame, exciting, Easy, challenging, Commonplace, new
HQ-Identification	isolating, integrating, amateurish, professional, gaudy, classy, cheap, valuable, non-inclusive, inclusive, unpresentable, presentable
Classic aesthetics	aesthetic, pleasant, clear, clean, symmetric, artistic
Expressive aesthetics	creative, fascinating, special effects, original, sophisticated

6.3.2 Iterative Open Coding

Next to existing domain-specific knowledge, the researcher may want to annotate the data using traditional coding practices (Strauss and Corbin, 1998). Coding consists of two steps. In *open coding* the researcher derives concepts from terms or phrases in the raw data. A concept may be defined through a direct relation to a specific term in the raw data (*in-vivo coding*; Glaser and Strauss (1967), for instance the researcher may create a concept termed *'friend'* referring to all terms *'friend'* in the raw data), or may relate to terms through an intermediate step of interpretation (for instance, the researcher might desire to group terms *'friend'*, *'family'* and *'colleague'* under the concept named *'relevant others'*). In a second step, termed *axial coding*, the researcher may define a hierarchical structure between concepts. The creation of this hierarchical structure is realized through the definition of a superordinate concept that includes all terms that appear in its subordinate concepts.

For instance, table 6.2 displays the full list of concepts along with coded examples, derived from the analysis of 329 narratives of the study described in chapter 4. Figure 6.1 displays a two-dimensional visualization of the similarity between all concepts including those deduced from domain-specific knowledge (table 6.1) and those derived from coding the data (table 6.2). Distances between concepts were derived from equation 6.7 and submitted to Multi-Dimensional Scaling. A two-dimensional solution was extracted. A-priori defined concepts are depicted in italics, while superordinate concepts are denoted in bold. Note that similar concepts (e.g. Novelty - Aesthetics in Interaction) often co-occur in the same narratives as displayed by the high similarity in the two-dimensional space.

One problem often met in coding procedures is the fixation of the researcher onto a certain perspective leading to increased biases in the interpretation process, what Kahneman et al. (1982) call anchoring bias. Strauss and Corbin (1998) proposed a number of techniques aimed at supporting the researcher in taking different perspec-

Table 6.2 Concepts that were derived from the analysis of the 329 narratives along with examples of coded data. Next to existing domain-specific knowledge, the researcher annotates data by relating individual terms or phrases to a set of latent concepts (linked concepts are denoted in [brackets]).

Concept	Individual terms
Self-expression	proud, show off, impress, status, jealous
Relevant others	friend, colleague, dad, daughter, brother
Sense of community	technology-freak, discuss, chat
Simplicity	easy, simple, straightforward, one click, seamless
Understandability	very logical, no clue how, i could not understand
Efficiency	almost instantly, no boot time, had to go through all the list
Effectiveness	organizes my messages nicely, simply not allowed, was not able to see
Fits context	use only one hand, wear gloves, while driving
Adaptable	adapts the screen brightness, totally customized, dynamically change
Familiarity	more used to, I am finding out, after using it for a while
Novelty	cool and intelligent design, genius idea, hay look at that its magic
Visual aesthetics	everything that shines, beautiful colors, shaky icons look beautiful
Aesthetics of interaction	glow effect, i just loved the way it turned around, so easy and fun
Fast access to information	immediate access, overhead of booting a computer, would start up my computer
Alleviate boredom in idle periods	in case I get bored, playing a bit every time, I can't really do anything else
Capturing momentary info	offload my own memory, have no means to note it down, added quite a text
Missing functionality	does not have, not supported, all phones can do it
Stimulation	[Novelty], [Visual aesthetics], [Aesthetics of interaction]
learnability	[simplicity], [Understandability]
Long-term usability	[Efficiency], [Effectiveness], [Fits context], [Adaptable], [Familiarity]
Useful	[fast access to information], [alleviate boredom in idle periods], [capturing momentary information], [missing functionality]

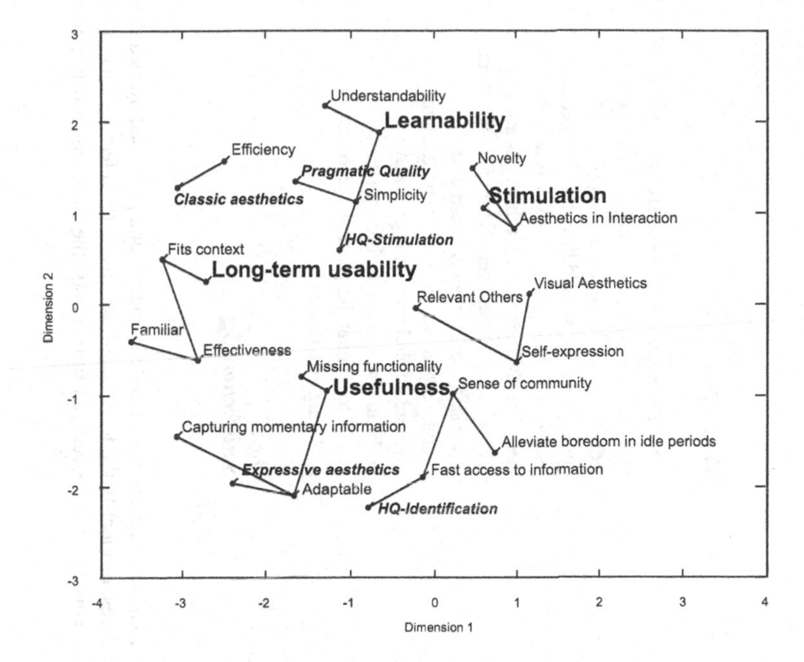

Fig. 6.1 Multi-Dimensional Scaling on concept dissimilarity (eq. 6.7). Concepts connected through lines belong to the same cluster. A-priori defined concepts are depicted in italics, while superordinate concepts are denoted in bold.

tives in the interpretation process. They argued that the very same comparison that can be performed on two objects (e.g. two flowers) resulting in the identification of the objects' properties (e.g. size, shape, color), may also be performed on theoretical concepts. For instance, the *flip-flop* technique involves a systematic comparison between a concept and its extreme opposite in identifying the concept's properties, e.g. what is meant by "immediate" access to information: time, effort or ubiquity.

Strauss and Corbin (1998) emphasize the role of diversity in the cases in the process of deriving new interpretations. This process may be supported through the use of interactive visualizations. For instance, figure 6.2 illustrates an interactive scatter plot that aims at enabling the researcher in browsing through diverse narratives. It computes the similarity between narratives from the co-occurrence of concepts (i.e. equation 6.6) and visualizes the first two dimensions derived from Multi-Dimensional Scaling. As more narratives are analyzed, the plot is updated. By selecting one of the points in the plot, the narrative is displayed on the right part of the interface and the coded data are highlighted and annotated by the respective concept. The researcher is further able to visualize other aspects of the experience narratives such as self-reported satisfaction for a given experience, day of occurrence for the reported experience, as well as the number of codes within each narrative.

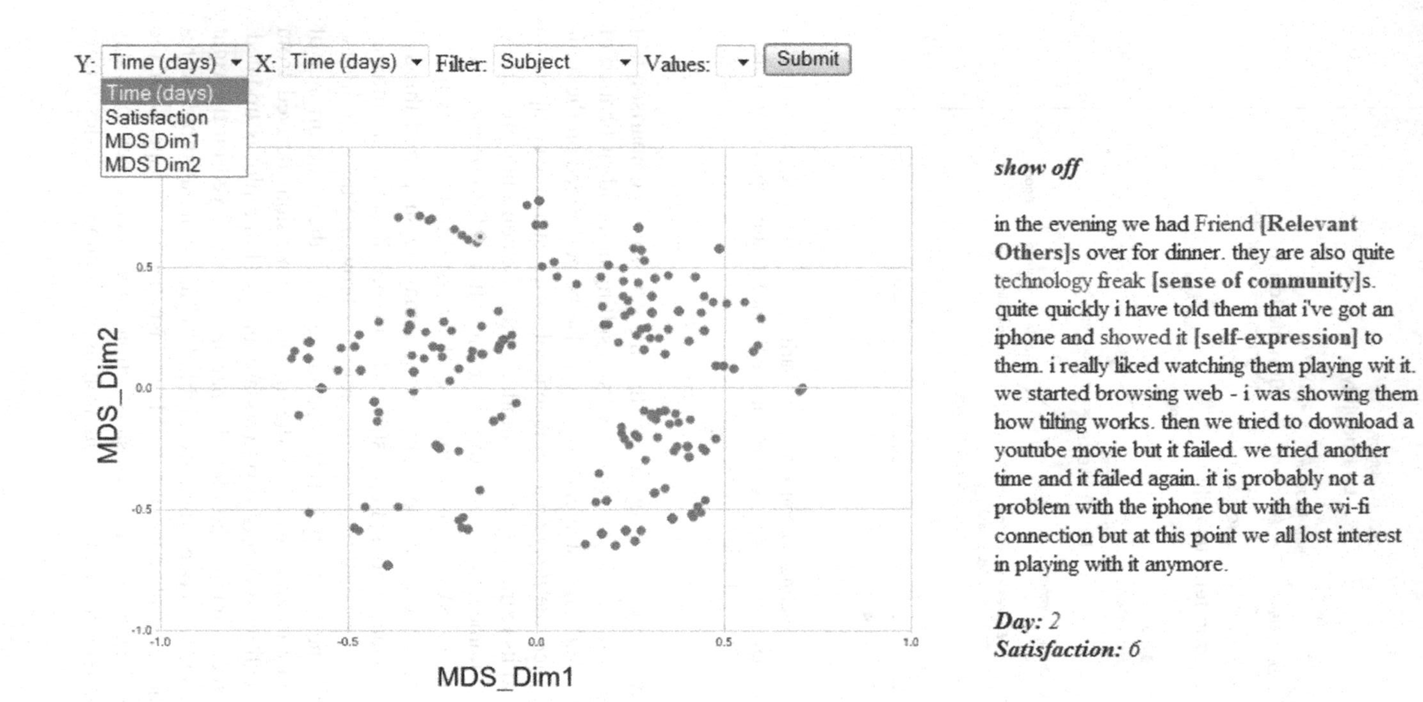

Y: Time (days) X: Time (days) Filter: Subject Values: Submit
Time (days)
Satisfaction
MDS Dim1
MDS Dim2
MDS_Dim2
MDS_Dim1
show off
in the evening we had Friend [Relevant Others]s over for dinner. they are also quite technology freak [sense of community]s. quite quickly i have told them that i've got an iphone and showed it [self-expression] to them. i really liked watching them playing wit it. we started browsing web - i was showing them how tilting works. then we tried to download a youtube movie but it failed. we tried another time and it failed again. it is probably not a problem with the iphone but with the wi-fi connection but at this point we all lost interest in playing with it anymore.
Day: 2
Satisfaction: 6

Fig. 6.2 Interactive scatter plot depicting the similarity between narratives derived from the co-occurrence of coded concepts and processed by means of Multi-Dimensional Scaling. The researcher may browse through different data instances, or identify the least coded narratives. By selecting a narrative the full text appears with the coded part highlighted and annotated by the respective concept. The analyst can further code the data, which will in turn influence the visualizations.

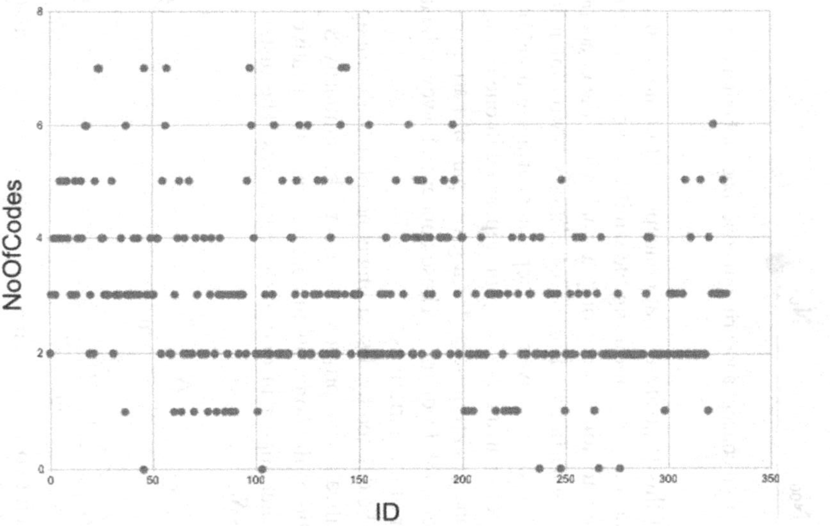
NoOfCodes
ID
0
2
4
6
8
0
50
100
150
200
250
300
350
Safari still doing great
my set-top box has some issues today so i could not check tv schedule. the alternative is on the internet but 1) i would need to go upstairs [long-term usability] [fits context] boot the computer and shout down to my wife [relevant others] to make a decision or 2) use iphone but i now the page is big Complicated [Pragmatic Quality] and has a long long drop-down [long-term usability] [efficiency] list of channels to pick. that list does not fit on any screen and is a pain even on a computer. surprisingly on iphone that long long drop-down list is implemented completely differently. in the keyboard area they show it in a completely custom way and the selection mechanism is even better than wit a regular browser on a computer. seriously cool easy [simplicity] and it even looks pretty!

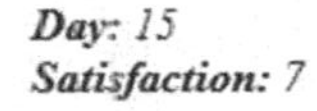
Day: 15
Satisfaction: 7

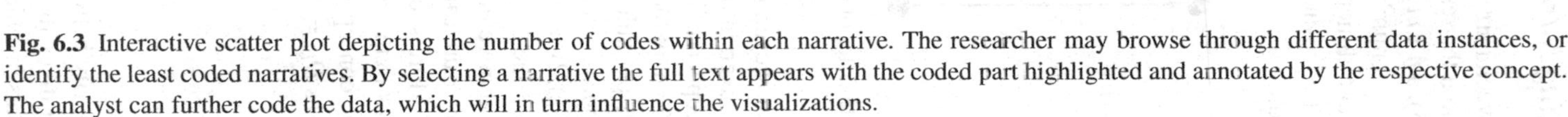

Fig. 6.3 Interactive scatter plot depicting the number of codes within each narrative. The researcher may browse through different data instances, or identify the least coded narratives. By selecting a narrative the full text appears with the coded part highlighted and annotated by the respective concept. The analyst can further code the data, which will in turn influence the visualizations.

6.3.3 Computing Narrative Similarity

The similarity between narratives is computed based on an explicitly defined latent space of concepts that the researcher is interested in. Using the vector-space model (Salton et al., 1975), a $n \times m$ matrix A is defined where each element $a_{i,j}$ depicts the presence or absence (or number of times) of concept i in narrative j. Each narrative is thus described by an n-dimensional vector and the similarity between the narratives can be equated to the cosine of the mutual angle between the two vectors, as in Latent-Semantic Analysis.

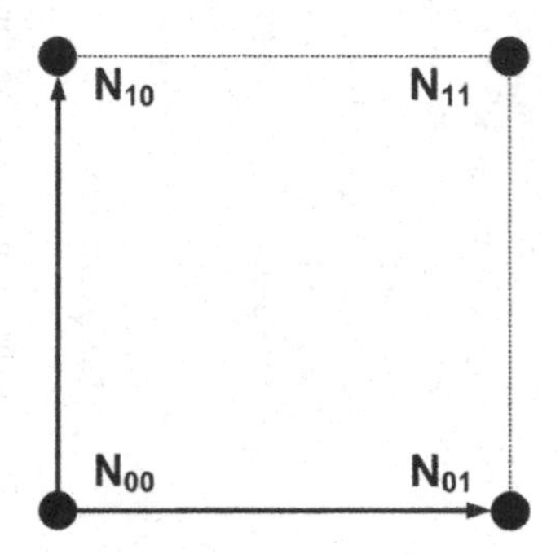

Fig. 6.4 Four possible states of a given concept appearing in none, one, or both narratives

One limitation, however, of this measure is that concepts that do not appear in both narratives are taken into account in computing the similarity between the narratives. These concepts are likely to dominate as only a small set of the concepts is likely to appear in 2 out of the 329 narratives. Figure 6.4 displays a two-dimensional configuration of all possible states for a given concept: appearing in none of the narratives (N_{00}), appearing in both narratives (N_{11}), or appearing in one of the narratives (N_{01} and N_{10}). If all concepts are plotted for a given pair of narratives, the N_{00} would dominate due to their high frequency. These concepts, however, bear no useful information about either of the narratives.

An alternative measure can be defined by discarding concepts that do not appear in either narrative. More specifically, we propose to equate the similarity $S_{i,j}$ between two narratives to the ratio of the concepts that appear in both narratives N_{11} over the sum of the concepts that appear in one of the narratives. The distance is then derived in 6.7 as $D_{i,j} = 1 - S_{i,j}$:

$$S_{i,j} = \frac{N_{11}}{N_{01} + N_{10}} \tag{6.6}$$

$$D_{i,j} = \frac{N_{01} + N_{10} - N_{11}}{N_{01} + N_{10}} \tag{6.7}$$

Likewise, the similarity between two concepts may be computed from the ratio of the number of narratives in which both concepts appear over the the sum of the number of documents in which one of the concepts appears.

6.3.4 *Hierarchical Clustering*

Once the similarity between all pairs of narratives has been computed, the distance matrix can be submitted to a clustering algorithm and the emerging clusters can be characterized by the three most dominant terms or concepts that appear in the narratives that are grouped together.

For instance, table 6.3 presents the three most dominant concepts for each of the first nine clusters that resulted from Ward's hierarchical clustering with a minimum variance criterium on the distance matrix of the 329 narratives. One may note that cluster 1 is dominated by experience narratives relating to Stimulation, clusters 4 and 8 mostly by narratives relating to learnability, clusters 3, 7 and 9 by narratives relating mostly to long-term usability, clusters 2, 5 by narratives relating to usefulness and narratives in cluster 6 relate to social experiences. This information may assist the researcher in further grouping the resulting clusters.

Table 6.3 Three most dominant concepts in each cluster

Cluster No.	terms.
Cluster 1	Stimulation, Novelty, Aesthetics in Interaction
Cluster 2	Missing functionality, Useful, HQ-Stimulation
Cluster 3	Long-term usability, Efficiency, HQ-Stimulation
Cluster 4	Learnability, Simplicity, HQ-Stimulation
Cluster 5	Useful, Fast access to info, HQ-Stimulation
Cluster 6	Relevant Others, Self-expression, HQ-Stimulation
Cluster 7	Long-term usability, Fits context, Adaptable
Cluster 8	Understandability, HQ-Stimulation, Pragmatic Quality
Cluster 9	Long-term usability, Effectiveness, simplicity

6.3.5 *Visualizing Insights*

Once narratives have been classified into categories that represent similar meanings, a number of generalized insights may be gained through the visualizations and exploration of the interrelations of meta-data such as the day in which a reported experience occurred (out of the 28 days of the study), and a self-reported satisfaction value (Osgood et al., 1957).

For instance, figure 6.5 illustrates a visualization of frequency and average impact of four types of experiences over the four weeks of the study. The Y axis corresponds to the number of experiences being reported for each respective experience type per week, while the average impact of experiences (as reported by the participants) for a given type are visualized through the size of each circle. Four overall types of experiences are distinguished: experiences relating to *stimulation, learnability, long-term usability* and *usefulness*. A 4-point (0-3) impact index of each reported experience is derived from the self-reported satisfaction elicited through a 7-point

(1-7) semantic differential scale 6.8. Satisfying experiences are differentiated from dissatisfying ones.

$$Impact_{0-3} = |Satisfaction_{1-7} - 4| \tag{6.8}$$

One may note that experiences related to learnability are the most dominant experiences during the first week. These experiences sharply decrease in the second week while the number of experiences relating to long-term usability and usefulness constantly increase over the four weeks of the study. Experiences related to stimulation seem to be the most impactful experiences during the first week followed by experiences of learnability. Surprisingly, experiences relating to long-term usability seem to be the least impactful despite their dominance over time. This finding did not become apparent in the initial analysis described in chapter 4 as the relation between time and impact of experiences was not explored. Visualization tools like the one illustrated in figure 6.5 enable rapid hypothesis construction and testing in the exploration of such data.

A different visualization, displayed in figure 6.6, can provide insights into the relations between *product qualities* (rows) and *product features* (columns, see table 6.4). Each cell in the matrix depicts the percentage of experiences referring in at least one instance to the respective product quality and product feature out of the total list of experience narratives that refer to the given feature (i.e. every column adds up to 100%). The researcher may for instance obtain insights into the overall experience of a given feature through the distribution of experiences over the five types,

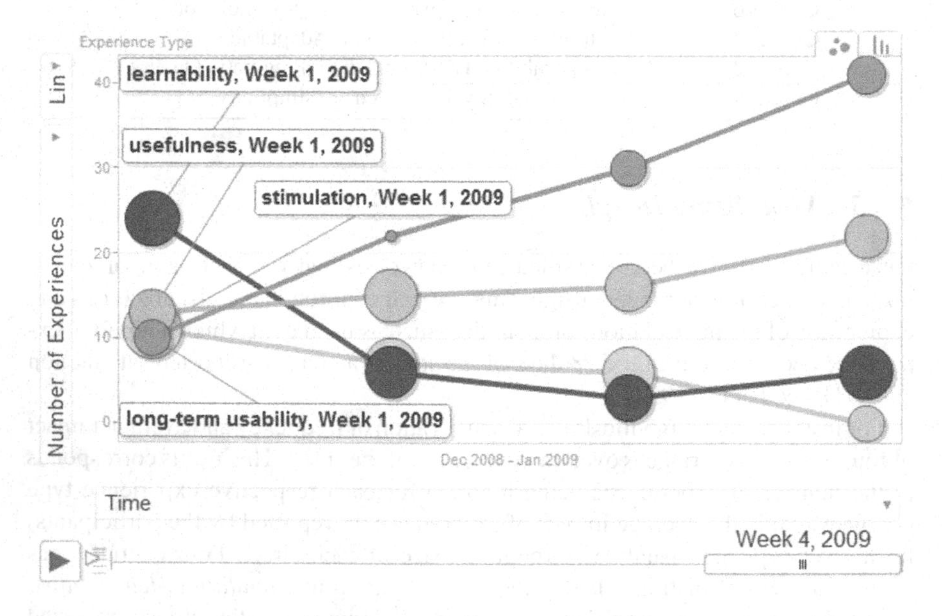

Fig. 6.5 Visualizing frequency of experiences (y axis) and average impact (size of circle) for four types of experiences colored differently over the four weeks of the study

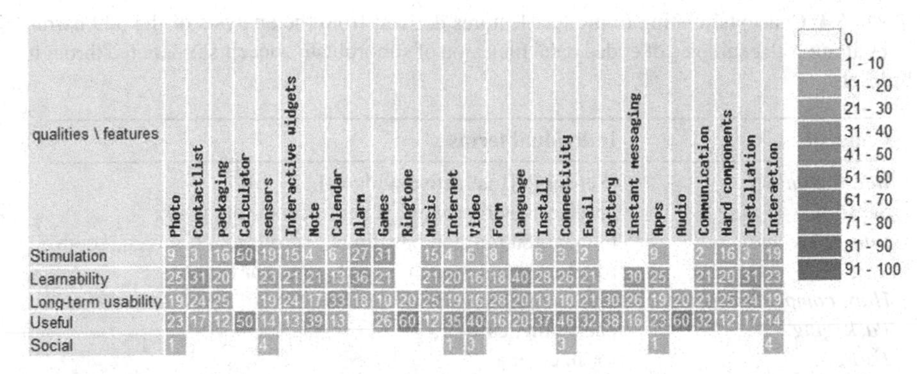

Fig. 6.6 Visualizing relations between product qualities and product features. Each column depicts the distribution of experiences over the five product qualities for a given product feature.

i.e. product qualities. This may lead, for instance, to the identification of the features that induce desired experiences such as the ones related to self-representation or the features that are dominated by learnability and long-term usability problems. By clicking on a cell, the full set of narratives relating to the respective product quality and product feature appear.

6.4 Validation of the Proposed Approach

Three distinct approaches were tested on a subset of the data in chapter 4. These were (a) Computing similarity on explicitly defined dimensions using the approach described in this chapter (Concept Analysis), (b) applying LSA on a restricted list of terms, and (c) applying LSA on all terms (traditional LSA). Thus, by splitting the proposed approach in two distinct procedures (a and b), we are able to distinguish between the impact of *i)* restricting the list of terms, and *ii)* defining explicit relations between concepts and the observed terms.

6.4.1 Preparing the Dataset

These amounted to a total of 347 experience narratives classified under five main categories: *stimulation, learnability, long-term usability, usefulness* and *social experiences*. Figure 6.7 displays the frequency and cumulative percentage of narratives across the number of words contained in each narrative. Large variances in the number of words across different narratives is likely to impact the computation of the semantic similarity of narratives with long narratives receiving higher similarity scores to other narratives as the probability of occurrence of words of interest is greater. This bias may be counterbalanced by weighting frequencies $a_{i,j}$ by the length of the document j. However, narratives below a certain number of words are not likely to contain adequate semantic information and may be excluded from

Table 6.4 Concepts relating to product features derived from the analysis of the 329 narratives along with examples of coded data. Inclusion of subordinate concepts is denoted through brackets.

Concept	Individual terms
Interaction	[Sensors], [Interactive widgets]
Sensors	accelerometer, proximity, sensor, rotation, guitar
Interactive widgets	interact, button, top bar, finger, scroll, touch screen, zoom, rotation, double-touch, sketch, draw, navigate
Hard components	[Packaging], [Form], [Battery]
Packaging	packaging, case
Form	form, color
Battery	battery
Interface	interface, menu, folder
Synchronization	sync, computer, laptop
Installation	[Language], [Install], [Connectivity], [Contactlist]
Language	language, english, dutch, dictionary
Install	install, configure, volume
Connectivity	WiFi, wi-fi, wifi, WIFI, wlan, wireless, network, bluetooth, blue tooth
Contactlist	contact
Apps	[Note], [Calendar], [Alarm], [Games], [Photo], [Calculator], [Internet], [Video], [Communication]
Note	note, todo, To-Do, to-do
Calendar	calendar, appointment
Alarm	alarm
Games	game
Photo	photo, picture, camera, webcam
Calculator	calculator
Internet	internet, safari, brows, web
Video	video, youtube, movie
Communication	[Instant messaging], [Email]
Instant messaging	sms, chat, messenger, facebook
Email	email
Audio	[Ringtone], [Music]
Ringtone	ringtone
Music	piano, equalizer, album, song, ipod, itunes

further analysis. For this dataset, narratives in the lower end of the graph (5%) with less than 24 words were excluded from further analysis, resulting in a total of 329 narratives.

6.4.2 Concept Analysis

The first approach used the procedure proposed in this chapter for the analysis of the 329 narratives. This analysis resulted in a total of 26 concepts referring to 539

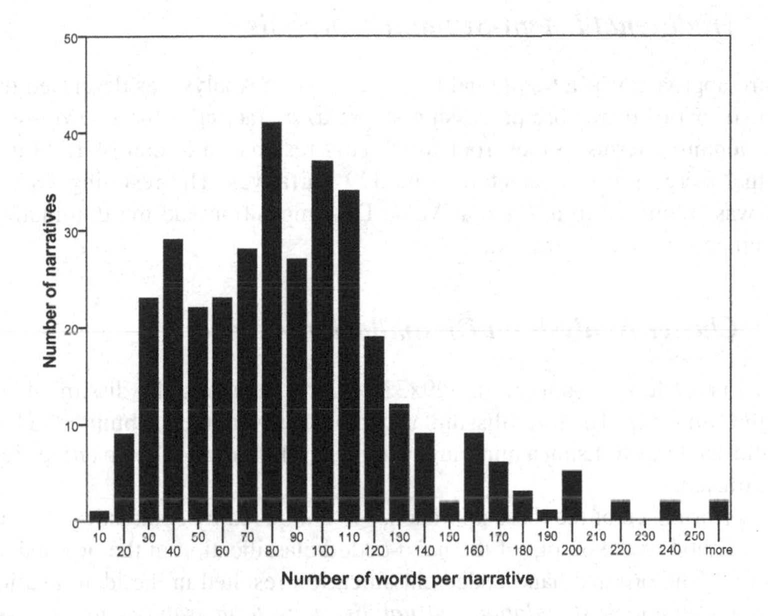

Fig. 6.7 Histogram on the length (number of words) of experience narratives. 5% of the narratives, i.e. the ones containing less than 24 words, were excluded from further analysis.

terms or phrases in the narratives. Five concepts were derived from existing domain-specific knowledge (see table 6.1) while the remaining 21 concepts were derived from the data through the qualitative coding procedure discussed in this chapter (see table 6.2.

The dissimilarity between narratives was then computed using the 26 concepts (equation 6.7), resulting in a 329x329 distance matrix.

6.4.3 Latent-Semantic Analysis on Restricted Terms

In the second approach, the explicit relations between concepts and terms were discarded. Instead, Latent-Semantic Analysis was applied using the restricted list of terms (539) that were identified by the researcher. Singular Value Decomposition was applied to the 539x329 matrix to extract the 26 most dominant latent dimensions. The optimal dimensionality in LSA is an ongoing research question, with some suggesting a dimensionality between 100 and 300 (Landauer and Dumais, 1997), while others suggest that most variance can be captured in the first 10 dimensions (Kontostathis, 2007). We applied a shared dimensionality of 26 in all three approaches to minimize any effects induced by differences in dimensionality when comparing the three approaches.

6.4.4 Traditional Latent-Semantic Analysis

The third approach was a traditional Latent-Semantic Analysis as described in this chapter. It involved two pre-processing steps: a) extracting a list of *stop-words* , and b) stemming terms to their root form. This resulted in a total of 1873 unique terms that were used to characterize the 329 narratives. The resulting 1873x329 matrix was submitted to a Singular-Value Decomposition and the dominating 26 latent dimensions were extracted.

6.4.5 Cluster Analysis on Dissimilarity Matrices

All three procedures resulted in a 329x329 matrix depicting the dissimilarity between the narratives. The three dissimilarity matrices were then submitted to hierarchical cluster analysis using a minimum variance criterium and the first *nine* clusters were extracted.

The performance of the three approaches is compared by contrasting the output of each method to the output of the hand-coded classification in the original study (chapter 4). The original hand-coded classification resulted in the identification of five overall categories: *stimulation, learnability, long-term usability, usefulness* and *social experiences*. Traditional content analysis, as applied in the original study, is considered as an optimal classification and used as reference for the three automated procedures.

To enable the comparison between the output of the three approaches with the output of the content analysis of the initial study, a mapping needs to be created between the 9 clusters generated by each of the three approaches and the five categories of the traditional content analysis. Once all 9 clusters are assigned to one of the five overall categories, interrater agreement indices such as the Kappa statistic (Fleiss et al., 2003), or the overall percent of correctly classified narratives may be computed in assessing the agreement between the three automated approaches and the traditional content analysis.

We employ two approaches for assigning each of the nine clusters to one of the five identified categories. First, this may be performed based on the distribution of narratives within a cluster over the five categories. The distribution for all nine clusters may be visualized in a *9x5* matrix where each cell of the matrix $m_{i,j}$ depicts the number of narratives that are classified to the cluster i (out of the 9 overall clusters that resulted from the automated analysis procedure) and to the j category (out of the 5 categories that resulted from the manual coding procedure in the initial study). According to this criterium, each cluster is assigned to that category that contains the highest number of narratives. This approach minimizes the error induced by the mapping process, and results in the best possible value for the agreement between the automated methods and traditional content analysis.

However, this best possible value may not be obtained in real settings where human interpretation is required to further classify the narratives. Thus, a second approach involves human raters. Each cluster, as proposed earlier in this chapter, can be characterized by the three most dominant terms in the experience narratives that are clustered together. This information may assist the researcher in interpreting the theme of each cluster and proceed in further grouping of the clusters.

Using the three most dominant terms for each cluster, six individuals were asked to assign each of the nine clusters to one of the five overall categories: *stimulation, learnability, long-term usability, usefulness* and *social experiences*.

Figure 6.8 displays the percentage of narratives being correctly classified, i.e. assigned to the same category as in the hand-coded analysis of the initial study Karapanos et al. (2009c), both for the optimal approach that employs the distribution of narratives across the five categories and for the approach that employs human raters in assigning the 9 clusters to five categories. The 95% confidence intervals depict the uncertainty derived from the human classification of the nine clusters into the five overall categories.

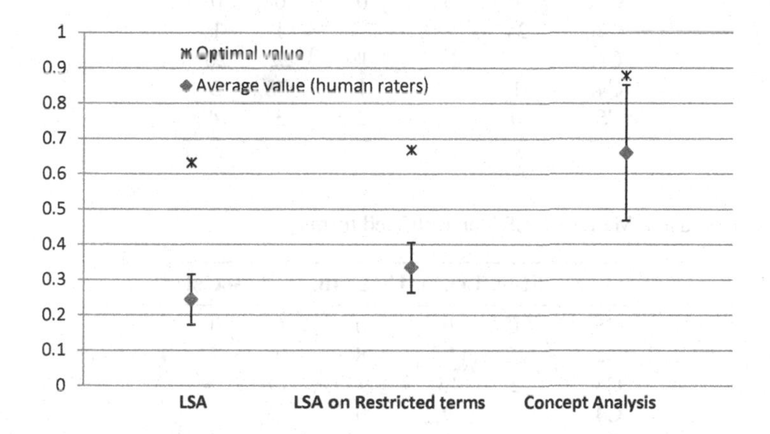

Fig. 6.8 Percentage of narratives being correctly classified, i.e. assigned to the same category as in the hand-coded analysis of the initial study Karapanos et al. (2009c), both for the optimal approach and for the approach that employs human raters in assigning the 9 clusters to five categories. The 95% confidence intervals depict the uncertainty derived from the human classification of the nine clusters into the five overall categories.

A number of insights are gained from this analysis. First, note the substantial difference in interrater agreement between the optimal mapping process (based on narratives distribution) and the process that involved human raters. This difference is larger in the two approaches that involved LSA as these procedures perform less favorably and are more ambiguous in characterizing the clusters through the three most dominant terms.

Next, the proposed method performs substantially better than traditional LSA, i.e. 66% of narratives were correctly classified versus 24% in the case of LSA when human raters are involved, while the optimal mapping process results in respective scores of 88% (Interrater agreement K=0.85 (Fleiss et al., 2003)) for the proposed method and 63% (Interrater agreement K=0.54) for traditional LSA.

Surprisingly, restricting the list of terms when applying LSA provides minimal improvement over the output of traditional LSA, i.e. on the full list of terms. Only when the explicit relations between terms and latent concepts are modeled there was a substantial improvement in the agreement with the hand-coded classification in the initial study.

Tables 6.5, 6.6 and 6.7 depict confusion matrices for the three semi-automated approaches (using the best possible value approach when mapping the nice clusters into five overall categories).

Table 6.5 Confusion Matrix for traditional LSA

	Stim.	Learn.	LT Usab.	Usef.	Social
C1.	0	0	0	0	0
C2.	24	51	23	1	1
C3.	1	23	79	22	20
C4.	1	1	1	78	2
C5.	0	1	2	3	4

Table 6.6 Confusion Matrix for LSA on restricted terms

	Stim.	Learn.	LT Usab.	Usef.	Social
C1.	0	0	0	0	0
C2.	22	53	8	10	7
C3.	2	13	95	19	15
C4.	3	9	0	72	1
C5.	0	0	0	0	0

Table 6.7 Confusion Matrix for Concept Analysis (proposed procedure)

	Stim.	Learn.	LT Usab.	Usef.	Social
C1.	24	0	0	0	0
C2.	41	67	3	6	4
C3.	0	0	89	0	0
C4.	2	8	11	90	0
C5.	0	0	0	5	19

6.5 Discussion

Overall, the proposed approach, was shown to display a substantially closer fit to the results of manual clustering of narratives in comparison to Latent-Semantic Analysis. However, interestingly enough, this was mainly rooted in the explicit modeling of relations between concepts and individual terms, and not in the restriction of the list of terms to the ones that concern particular phenomena of interest.

At the same time, the proposed approach displays a number of advantages over traditional manual coding procedures as concepts already defined through the analysis of a subset of narratives, are indexed and used to characterize all narratives. This has two main implications. First, in the case of an established body of knowledge in the domain, the researcher can start from an initial classification. Second, the researcher is assisted by the iterative process of coding and visualization where the impact of newly defined concepts is directly visible on the full set of data.

This, however, also introduces some complications for the coding process. When defining a concept through the interpretation of a certain term or phrase, the researcher is also forced to assess the generalizability of the relation between term and concept under different contexts. In principle, the researcher needs to avoid coding lengthy phrases as the likelihood of occurrence of the same phrase in other narratives decreases as its length increases. On the other hand, a narrow interpretation of a single term in a given context entails the risk of false attribution of meaning of the term in other contexts. Thus the researcher needs to find a balance between, on the one hand, effective characterization of narratives, and on the other hand, omitted risks of false interpretation in other contexts. We argue that this process minimizes the risks of over-interpretation of qualitative data as the researcher is forced towards a systematic comparison of the use of a given term or phrase across different contexts.

While the proposed approach argues for the aggregation of experiences into generalized knowledge, it does not intend to downplay the role of idiosyncratic insights in the design process. Instead, it argues for an interplay between generalized knowledge and idiosyncratic insights. The latter ones may provide a rich understanding on how the product is experienced in a given physical and social context. They may lead to an understanding of users' needs and generate ideas for design. But, design needs prioritization. Generalized knowledge may point at the important issues. For instance, how frequent is a given experience? What experiences dominate over prolonged use? Next, not all experiences are equally interesting to the designers. An information-theoretic approach would suggest that interestingness relates to the rarity of an experience. Thus, by visualizing the interrelation between experience narratives (see fig. 6.2) we enable the researcher to identify experiences that bear limited similarity to others and are thus more likely to contain new information for design.

The proposed approach is only a first step towards more automated approaches for visualizing and interacting with qualitative data. A number of limitations and future directions may be noted. First, coding is now restricted to matching terms or phrases whereas more advanced coding procedures might also include logical operations. For instance, the researcher might desire to identify a concept when two terms appear (at any place) within the narrative or when a term appears while another term is absent (e.g. the term *beautiful* is likely to relate to the concept of *visual aesthetics* only when terms like *simple* and *interactive* are absent). Similarly, axial coding is now restricted to establishing a hierarchical structure between the concepts while a wider spectrum of relations would enable a richer exploration of data. For instance, causal relations between concepts might be explicitly identified by the researcher or automatically established by the relative position of concepts within the unit of analysis.

6.6 Conclusion

This chapter highlighted two problems in the qualitative analysis of experience narratives. First, qualitative analysis is a labor intensive activity which becomes increasingly a concern when qualitative data may be elicited from a large amount of participants as in the case of iScale. Second, qualitative analysis has been shown to be prone to researcher bias as humans often rely on heuristics in forming judgments about the relevance or similarity of two or more data instances (Kahneman et al., 1982).

This chapter proposed a semi-automated approach that aims at supporting the researcher in the content analysis of experience narratives. This approach relies on a combination of traditional qualitative coding procedures (Strauss and Corbin, 1998) with computational approaches to the assessment of semantic similarity of documents (Salton et al., 1975). The approach shares a number of advantages over traditional content analysis procedures as the coding scheme derived in the analysis of a small set of data is used to characterize all remaining data. Thus, through an iterative process of coding and visualization of insights, the approach enables the researcher in moving between highly idiosyncratic insights and generalized knowledged. Secondly, as the researcher is forced to examine the use of language under different contexts, it minimizes the risk of over-interpretation which is common in traditional qualitative analysis practices.

Using data from chapter 4, the performance of the proposed approach was compared to the one of a fully automated semantic analysis procedure, the Latent-Semantic Analysis Deerwester et al. (1990). The proposed approach was found to display a substantially closer fit to the results of manual clustering of narratives in comparison to Latent-Semantic Analysis.

Chapter 7
Conclusions

The focus of Human-Computer Interaction has evolved from the study of the usability of interactive products towards a more holistic understanding of the psychological and social impact of products in people's lives. Firstly, this has resulted in a wealth of new concepts such as pleasure (Jordan, 2000), fun (Blythe et al., 2003), aesthetics (Tractinsky et al., 2000), and hedonic qualities in the use of personal interactive products (Hassenzahl, 2004), but also aspects of trust in online transactions (Egger, 2003), and the increased social connectedness that awareness systems bring among family members (IJsselsteijn et al., 2009; Markopoulos et al., 2004; Van Bel et al., 2008). Secondly, it has lead to an increased emphasis on methods for assessing the subjective quality and psychological consequences of product use. While a wealth of methods and techniques are available for assessing the usability of interactive products, research on methods for the subjective assessment of users' experiences is only at its infancy (e.g. Van Schaik and Ling, 2007; Fenko et al., 2009; Ben-Bassat et al., 2006; Zimmerman et al., 2009; Väänänen-Vainio-Mattila et al., 2008).

7.1 Contributions of This Work

In this work we identified the notion of diversity in subjective judgments of the quality of interactive products. We identified two distinct schools of thought in the field of user experience and highlighted the methodological issues they pose when one is concerned about capturing the diversity in users' responses to interactive products. More specifically, the focus of our work has been threefold:

- to conceptualize the notion of diversity in subjective judgments of users' experiences with interactive products
- to establish empirical evidence for the prevalence of diversity, and
- to provide a number of methodological tools for the study of diversity in the context of product development.

E. Karapanos: Modeling Users' Experiences with Interact. Syst., SCI 436, pp. 137–146.
springerlink.com

7.1.1 Conceptualizing Diversity in User Experience

A long-standing assumption in the field of psychophysics is that different individuals will more or less agree on perceptual judgments such as how much noise, or blur, an image contains, or how much friction, or inertia, one may find in a haptic control. This work highlighted that this assumption, often referred to as the *principle of homogeneity of perception* (Martens, 2003), does not necessarily hold in the context of cognitive judgments of the quality of interactive products.

Using a framework from (Hassenzahl, 2005), we conceptualized diversity as existing at two different stages in the formation of an overall evaluative judgment about an interactive product. *Perceptual diversity* lies in the process of forming product quality perceptions (e.g. novel, easy to use) on the basis of product features. For instance, different individuals may infer different levels on a given quality of the same product, e.g. disagree on its novelty. *Evaluative diversity* lies in the process of forming overall evaluations of the product (e.g. good-bad) on the basis of product quality perceptions. For instance, different individuals may form different evaluative judgments even while having no disagreement on the perceived quality of the product, e.g. both might think of it as a novel and hard-to-use product, but they disagree on the relative importance of each quality.

7.1.2 Establishing Empirical Evidence for the Prevalence of Diversity in User Experience

We identified two critical sources for diversity in the context of users' experiences with interactive products: (a) interpersonal diversity in users' responses to early conceptual designs, and (b) dynamics of users' experiences over time.

7.1.2.1 Interpersonal Diversity in Users' Responses to Early Conceptual Designs

Den Ouden (2006) highlighted that design decisions made early in the conceptual phase and particularly the ones relating to product definition are responsible for the majority of soft reliability problems. This leads to questioning the degree to which such design decisions are grounded on empirical insights about users' preferences. Based on a number of informal communications with stakeholders in concept design practices, we observed that it is often difficult to trace back the reasons that motivated certain design decisions. The questions raised then were on what basis are design decisions made? Can designers really foresee users' preferences? Chapter 2 presented a study that inquired into the differences between designers' and users' views on a set of early conceptual designs. The study highlighted that designers' views can be substantially different from the ones of designers. Even in a case where designers and users preferred the same product, users were found to base their preference on substantially different reasons, i.e. they valued different product qualities than the ones valued by the designers. It was further highlighted that both designers and users, in judging the quality of these conceptual designs, elicited a wealth of personal attributes that could not be captured by standardized questionnaires. Chapter 2 highlighted that diversity also exists within a single individual, in the sense that different attribute judgments of a participant may reveal different, complementary, views. This diversity in perspective cannot be represented in a single averaged view but requires multiple diverse views.

7.1.2.2 Dynamics of Users' Experiences Over Time

Product evaluation practices have traditionally been focusing on early interactions and product qualities that are salient in these interactions. This work highlighted that diversity exists not only across different individuals but also across time, in the sense that different product qualities are valued at different phases in the adoption of the product. Chapter 4 presented two studies that inquired into the dynamics of users' experiences with interactive products. In the first study, judgments of the overall *goodness* of a novel pointing device were found to shift from a use-based evaluation dominated by the pragmatic quality of the product, i.e. usefulness and ease-of-use, to an ownership-based evaluation dominated by aspects of identification, i.e. what the product expressed about their self-identify in social contexts. Judgments of *beauty* seemed to relate to perceptions of novelty during initial interactions, but this effect disappeared after four weeks of use. The second study followed six individuals through an actual purchase of an Apple iPhone and inquired into how their expectations and experiences developed one week before and four weeks after the purchase of the product. The study revealed that the product qualities that provided positive initial experiences were less crucial for motivating prolonged use. A tentative framework was proposed that identifies three distinct phases in the adoption of a product: *orientation, incorporation* and *identification*, each referring to a different relation between user and product, with different qualities being most salient.

7.1.3 Proposing Methodological Tools for the Study of Diversity

Lastly, we proposed a number of methodological tools for the study of the two different sources of diversity: (a) understanding interpersonal diversity through personal attribute judgments, and (b) understanding the dynamics of experience through experience narratives.

7.1.3.1 Understanding Interpersonal Diversity through Personal Attribute Judgments

Traditional approaches to measuring users' responses to artifacts derived through parallel design (Nielsen and Desurvire, 1993) lie in the use of validated psychometric scales where measures are being defined a-priori by the researchers. We identified two limitations of such practices when one is concerned about inquiring into the diverse ways with which users form evaluative judgments of interactive products. First, it was argued that the a-priori definition of relevant measures is inherently limited as researchers might fail to consider a given dimension as relevant for the given product and context, or they might simply lack validated measurement scales for a relevant dimension. Secondly, such approaches assume that participants are able to interpret correctly and position a given statement, that is defined by the researcher, in their context. Recent literature has challenged this assumption, suggesting that in certain cases participants are unable to interpret the personal relevance of the statement in their own context; instead, they employ *shallow processing* (Sanford et al., 2006), i.e. responding to surface features of the language rather than attaching personal relevance to the question.

We argued for an alternative approach that lies in a combination of structured interviewing, that aims at eliciting the attributes that are personally meaningful for each individual, with a subsequent rating process performed on the attributes that were elicited during the interview. The Repertory Grid Technique was proposed as a promising attribute elicitation technique as it aligns well with parallel design practices that are typically met in the concept design phase (Nielsen and Desurvire, 1993; Hassenzahl and Wessler, 2000).

Despite the promising nature of the Repertory Grid Technique, it was argued that current analysis procedures are not suited for interpreting such data. Two dominant approaches in the analysis of repertory grid data were identified: a) a qualitative, idiosyncratic approach with a narrative summarization, and b) a quantitative approach that employs averaging procedures using exploratory multivariate techniques. We argued that these two approaches are limited in two respects. Qualitative approaches do not take properly into account the elicited attribute ratings and thus do not fully exploit the true value of the Repertory Grid Technique, which is to quantify rich qualitative insights. Averaging procedures, on the other hand, treat diversity among participants as error and thereby contradict the basic idea of The Repertory Grid and relevant personal attribute elicitation techniques.

Thus the lack of appropriate statistical analysis techniques was established and two Multi-Dimensional Scaling procedures that aim at quantifying the diversity of repertory grid data were proposed. The inadequacy of traditional averaging practices was established as in two case studies they were found to account for 15% and 43% of the available attributes only; the remaining attributes had minimal contribution to the configuration space and thus could not be adequately modeled by the average model. The technique proposed in chapter 3 was found to account for more than double of the attributes accounted for by the average model, to provide a better model fit even for the attributes that were adequately predicted by the average model, and to result in semantically richer insights, since the diverse models can account for more semantically different attributes.

7.1.3.2 Understanding the Dynamics of Experience through Experience Narratives

Traditional approaches to measuring the dynamics of experience over time typically employ validated measurement and structural models across different phases in the adoption of a system (e.g. Venkatesh and Davis, 2000; Venkatesh and Johnson, 2002; Kim and Malhotra, 2005). We highlighted a number of limitations of such approaches. First, they rely on an assumption that the relevant latent constructs remain constant, but their perceived value and relative dominance change over time. As it was shown in chapter 4, however, prolonged use might relate to a different set of user experiences such as *daily rituals* and *personalization* that do not become apparent in users' initial interactions and that may thus be not captured by the measurement model. Reversely, other constructs that are salient in initial use situations may become irrelevant over prolonged use. This may lead to distorted data, for instance when participants are not able to interpret the personal relevance of a given measure (e.g. learnability) in the current usage situation. Last, such approaches provide rather limited insight into the exact reasons for changes in users' experiences. They may, for instance, reveal a shift in the dominance of perceived ease-of-use and perceived usefulness on intention to use a product (e.g. Venkatesh and Davis, 2000), but provide limited insight to the exact experiences that contributed to such changes. Insights into these rich and contextualized data is what designers need in designing for a given context.

We argued for an alternative approach to the measurement of the dynamics of experience over time, that relies on a) the elicitation of idiosyncratic self-reports of one's experiences with a product, the so-called *experience narratives*, and b) the extraction of generalized knowledge from a pool of experience narratives through content analytical procedures. In this sense, each narrative may provide rich insights into a given experience and the context in which it takes place. However, generalized knowledge may also be gained from these experience narratives. Such generalized knowledge may be reflected in questions like: how frequent is a certain kind of experience, what is the ratio of positive versus negative experiences and how does this compare to competitive products, how does the dominance of different product qualities fluctuate over time and what should we improve to motivate prolonged use?

This leads to two research problems. Firstly, how can we elicit experience narratives efficiently? Chapter 5 reviewed existing methodological paradigms for inquiring into users' experiences over time. Longitudinal designs such as the one employed in chapter 4 were identified as the gold standard in the study of long-term effects of product use, but their labor-intensive nature was highlighted as a barrier towards their adoption but also as an antecedent of restricted samples in terms of the user population, the product population and studied time. An alternative methodological approach was proposed that relies on the elicitation of user' experiences with a product from memory. The chapter presented iScale, a tool that was designed with the aim of increasing participants' effectiveness in recalling their experiences with a product. Two different versions of iScale, the *Constructive* and the *Value-Account* were motivated by two distinct theoretical approaches in the reconstruction of one's emotional experiences. These two versions were tested in two separate studies against a control condition, free-recall employing no process of enhancing users' recall. Overall, iScale was found to result in a) an increase in the number of experience reports that participants provided, b) an increase in the amount of contextual information for the reported experiences, and c) an increase in participants' accuracy in recalling concrete details of the experienced events, thus suggesting that iScale is able to minimize retrospection biases when recalling one's past experiences with a product.

Secondly, how can we aggregate the idiosyncratic experiences into generalized knowledge? Chapter 4 presented a case where content-analysis was employed in deriving key themes in the data, classifying narratives into a set of main categories and identifying the distinct distributions over time across the different categories. Two limitations were identified in this procedure. First, it is a labor intensive activity which becomes increasingly a concern when qualitative data may be elicited from a large amount of participants as in the case of iScale. Second, it is prone to researcher bias as humans often rely on heuristics in forming judgments about the relevance or similarity of two or more data instances (Kahneman et al., 1982). Chapter 6 proposed a novel technique for the semi-automated analysis of experience narratives that combines traditional qualitative coding procedures (Strauss and Corbin, 1998) with computational approaches for assessing the semantic similarity between documents (Salton et al., 1975). It was argued that the proposed approach supports the researcher through semi-automating the process of qualitative coding, but also minimizes the risks of overemphasizing interesting, but rare experiences that do not represent users' typical reactions to a product.

7.2 Implications for the Product Creation Process

We argued against the distinction between *evaluative* and *inspirational* purposes of product evaluation which constitute the current practice (figure 7.1). Two methods were proposed that aim at extrapolating rich qualitative information and informing the validation process. This leads to two benefits in the evaluation of interactive products. Firstly, it enhances the content validity of the validation process as the measurement model reflects more accurately the dominant dimensions of users' experiences with the product. Secondly, it enhances the validity of the extrapolation process as it quantifies the significance of given experiences and as such it minimizes the risks of overemphasizing interesting but rare insights.

7.2.1 Integrating Subjective and Behavioral Data

This work focused on methods for the subjective evaluation of interactive products. How can such methods be combined with objective information regarding users' behavior? Behavioral data may not only provide insights into the actual usage of a product but may also inform the elicitation of users' experience. In Funk et al. (2010) we provided a case study where usage information was used to augment subjective feedback in two ways (figure 7.2). First, users were able to provide subjective feedback at the times they liked as in traditional event-based diaries (Bolger et al., 2003). Through process mining techniques (van der Aalst et al., 2007) we were able to inquire into users' interactions that preceded the users' reports. Second, through an observation specification language (Funk et al., 2008) we were able to identify interaction patterns that we were interested in and probe for subjective

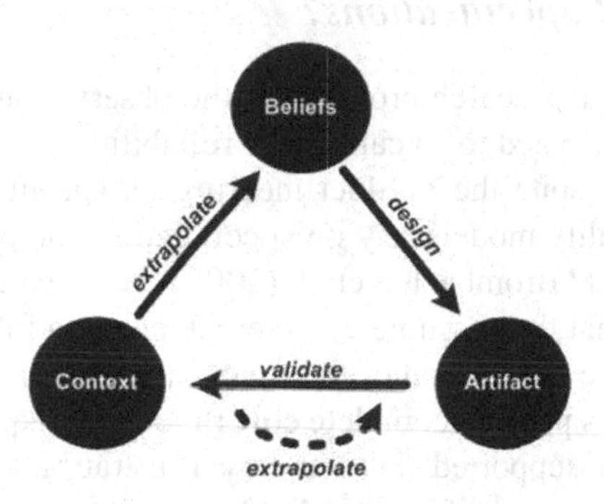

Fig. 7.1 A simplified framework of the design process, derived from (Martens, 2009b, personal communication). According to this framework, the designer forms beliefs about the potential consequences of use of a given product in a given context, grounded on empirical knowledge. These beliefs are in turn externalized to artifacts and the artifacts are validated on a number of criteria that have been derived from prior beliefs, or empirical insights from the previous design iteration.

feedback as in the experience sampling method (Hektner et al., 2007). In this way, we were able to identify two distinct usage modes of a media recommender system, i.e. goal-oriented interaction as in the case of searching for specific information and action-oriented interaction as in the case of browsing through media content (Hassenzahl, 2005).

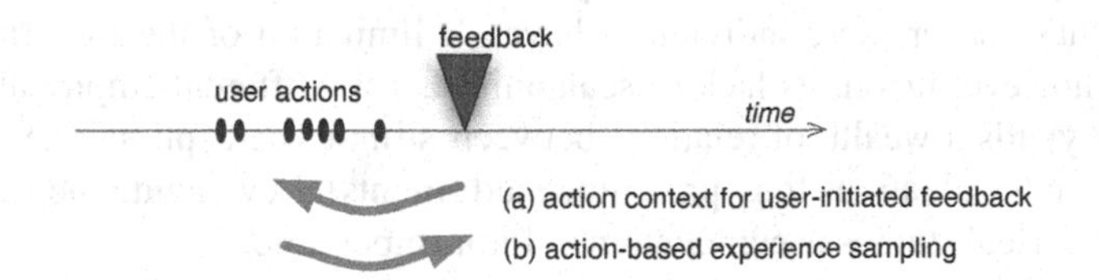

Fig. 7.2 Information about preceding user actions can (a) provide the context for user-initiated feedback, and (b) trigger event-based experience sampling.

This approach enabled us in capturing momentary experiences that pertained to interactions that we were interested in. We observed however that experience sampling is often intrusive to users' interactions and can lead to loss of information as participants often cannot interrupt the activity they are engaged in to provide subjective feedback regarding their experience (see also Hsieh et al., 2008). One could thus integrate such in-situ field feedback with methods for retrospective feedback as in the case of iScale, or the Day Reconstruction Method (Khan et al., 2008, for instance, see). These methods will be attempted in UXSuite (Koca and Funk, 2009), a start-up company supported by a STW valorization grant.

7.2.2 *The End of Specifications?*

The starting point of this research project was the observation that about half of returned products are attributed to so called soft reliability problems, i.e. cases where the user complaints despite the product meeting its specifications. The question raised then is, can quality models rely on specification compliance as a metric for the quality of a product? Brombacher et al. (2005) criticized traditional quality and reliability models in that they assume a) the existence of a full set of specifications that provide a complete description of the functional behavior of a product, and b) that these specifications provide complete coverage of all requirements of the product. This work further supported this view by illustrating that user requirements may substantially vary across different individuals and that these requirements may develop over time as the users appropriate the product in their own context.

7.3 Avenues for Future Research

We proposed a number of methods for quantifying the diversity in users' experiences with interactive products. A number of limitations may however be noted, leading to interesting avenues for future research.

7.3.1 *Leveraging Insights across Different Exploratory Studies*

One of the advantages of the Repertory Grid Technique is that it provides rich insights into individuals' idiosyncratic views on the set of stimuli. We argued for exploratory quantitative analysis techniques that can adequately model the different views that one or more individuals have. A limitation of the Repertory Grid Technique however lies in its lack of scalability across different empirical studies. Each study yields a wealth of relations between stimuli, perceptions (i.e. personal constructs) and evaluations (e.g. preference judgments). Leveraging the insights of different empirical studies is cumbersome, if not impossible.

Structured interview techniques, however, such as triading (Kelly, 1955), laddering (Reynolds and Gutman, 1988) and pyramiding (Fransella et al., 2003), that are typically employed in repertory grid interviews have the advantage of imposing a set of relations between the constructs that are being elicited, and are thus *computational friendly*. Empirical insights could thus be modeled through a graph $G=(V,E)$ where the vertices V reflect the identified constructs and the edges E reflect the empirically identified relations between the constructs, and each edge, i.e. a relation between two constructs, could be weighted by its frequency of appearance in empirical studies. This might lead to a new set of tools that could have a substantial impact on science and design.

7.3.2 Computational Tools for Making Survey Research Scalable

Psychometric scales can be characterized as one of the most ubiquitous measurement tools in the social sciences. However, they are not free of problems. The development of a set of scales is often described as a three-step process: item generation, scale development, and scale evaluation (Hinkin, 1995). The first step aims at enhancing the content validity of the questionnaire (i.e. that a complete coverage of the domain of interest is obtained through the proposed items); the latter two steps aim at enhancing the convergent and discriminant validity of the questionnaire (i.e. that each item correlates highly with other items that attempt to measure the same latent construct, and weakly with items that attempt to measure different latent constructs).

While the later two phases are supported by a wealth of statistical techniques, the item generation phase (related to the content validity of the questionnaire) is still regarded as a largely subjective procedure (Scandura and Williams, 2000; Hinkin, 1995; Larsen et al., 2008b). Questionnaire items are typically generated through brainstorming with domain experts, or through empirical studies with participants from the targeted audience, most often involving structured interviewing techniques (c.f. Hinkin, 1995). A number of limitations can be identified in item generation practices.

First, brainstorming practices require a firm understanding of the problem domain and clear definitions of the constructs to be measured. However, as Haynes et al. (1995) noted, questionnaires are often grounded in contemporaneous theories that evolve over time and that are supported by limited empirical data, thus, it is inevitable that early questionnaires fail to capture all possible facets of a construct and consecutive iterations are required for an adequate measurement of the construct.

Second, domain experts in brainstorming item generation practices often resort to lexical similarity (i.e. synonymy) when deriving new items in an effort to assure high convergent validity of the proposed constructs. This may have substantial implications to the rating process. Larsen et al. (2008b) found for the majority of constructs in a sample of questionnaires, the semantic similarity between items to be a significant predictor of participants' ratings ($.00<R^2<.63$). In such cases, participants are more likely to have employed shallow processing (Sanford et al., 2006), i.e. responding to surface features of the language rather than attaching personal relevance to the question.

Ideally, questionnaire items should be grounded on a large pool of empirical studies that are likely to have identified multiple facets of a given construct. Constructs could then evolve as new knowledge is being added to the corpus. Such a practice is however currently infeasible due to a lack of scalability across different empirical studies, as the exact procedure of item generation, the full-list of generated items, and their relationships are frequently not properly reported (Hinkin, 1995; Scandura and Williams, 2000). A computational infrastructure such as the one sketched above would enable social scientists in leveraging insights across different exploratory empirical studies which would in turn lead to scalable scale development practices within and across different research groups.

7.3.3 Empirical Knowledge Bases for Forming Design Goals

Designers form beliefs. In the absence of empirical insights, designers base these beliefs on prior experiences and their own intuition. Empirical knowledge derived from similar artifacts in similar contexts may not only make these beliefs grounded on actual evidence, but may also inspire the process of forming beliefs. Computational infrastructures such as the ones described above may provide rich insights into the relationships between design parameters, contextual parameters, and their psychological consequences (e.g. product quality perceptions).

7.3.4 A New Basis for User Insights?

The computational infrastructure proposed above assumes the use of structured interview techniques such triading, laddering and pyramiding; these share the benefit of imposing a set of relations between the constructs being elicited and thus are more "computational-friendly". The iScale tool lead to a new challenge rooted in providing access to immense, unstructured qualitative data. Chapter 6 proposed a computational approach for the content analysis of unstructured qualitative data such as in the case of experience narratives. One may however note that content analysis is limited in its scope as it looks only for one type of relation, that of similarity/dissimilarity between different data instances (e.g. narratives). The question then becomes: how can computational tools assists researchers in identifying a richer set of relationships such as causal effects in unstructured qualitative data? Can this procedure be partly automated, e.g. uncovering causal relationships through the relative position of concepts within the unit of analysis? If not, how can we assist the researcher in transforming qualitative data to a computational format (i.e. a graph network) that would enable him in rapid hypothesis testing, e.g. counterfactual analysis (King et al., 1994) within her own dataset as well as across different datasets?

Next, given the access to computational tools that partly automate the analysis of rich qualitative data and support the identification of emerging patterns, we foresee a new basis for user insights, one derived from internet forums, complaint centers, emails and other sources of public statements. These may lead to user insights relating to product success in the market but also to emerging trends and unidentified needs in the marketplace. With this work, through the development of the iScale tool and the Computational Content Analysis method, we aimed at making a first step in inquiring into richer sources of user insight data.

Closing Note: We Are about Time, We Are about Change

Since I conducted my very first usability test back in 1996, I was intrigued by how much happens during this hour of testing. Difficult barriers just disappeared after a while. At the same time, new problems and challenges crept up. Participants learned, assimilated, accommodated and even already appropriated the tested piece of technology. Consequently, impressions and opinions of participants after the test did not always easily match their opinions and impressions just 10 minutes into the test.

My clients however expected an overall assessment of the product: How usable is it and what do we need to do? So I busily averaged task completion rates and times, counted and collected critical incidents, and interpreted post-test questionnaires. But it made me feel uneasy. Isn't it rather the change over time, its shape, its speed, which I should look at? Instead, I averaged it all out!

Now, some people may not consider an hour of usability testing as longitudinal. It clearly is, but there is also a clear difference between a time span of hours and weeks. I got interested in a larger scope. Back in 2005 together with Christiane Hartmann and Sanna Cavar I set out to explore what could possibly be learned from a longitudinal perspective on a product (see also Hassenzahl, 2010, pp. 22). We gave eight participants a version of Will Wrights classic social simulation *The Sims - livin large*. The participants never played a computer game before. Their task was to play at least 15 minutes a day for a period of 28 days. We gave them a diary for each day to record thoughts, feelings and behaviour. It started with asking about their momentary affective state (before playing) and the time they plan to spend with the game on that day. While playing, they had "event cards" to report those events, which clearly changed their feelings toward the game. They noted what happened (e.g., "my Sim got a job") and rated their feelings toward these events. After the games session, they were again asked to rate their momentary affective state, how much they liked *Sims* and to note how long they played on this day. And oh my the *Sims* are tempting: participants played on average 19 minutes more then they planned for, 34 min versus 53 min, $t(27)=-8.69$, $p<.001$.

Figure 1, lower part "frequency" shows the mean number of reported events for each day. For example, on day 1 this is 1.5, representing a total of 12 reported events by 8 participants. This number is declining over time (see the gray line, $r= -0.56$,

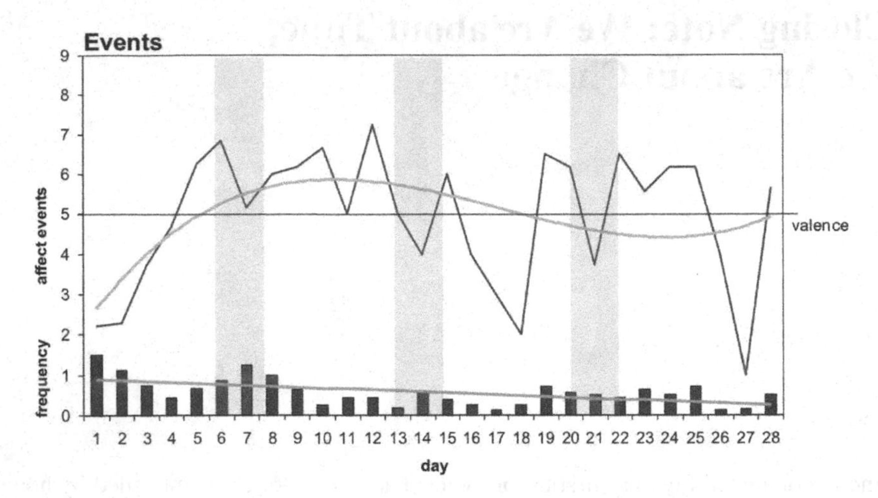

Fig. 7.3 Events

p<.01). The longer participants played, the smaller the number of events reported. More interesting, however, is the mean valence of the events, that is how positive or negative the events had been on each day (Figure 1, upper part "affect events"). The scale runs from 1 "negative" to 9 "positive," with 5 being "neutral." Interestingly, it needed four days of play, before the events became positive. On day 1 participants had problems with installing the game. On day 2 they struggled with the controls, overview and the pace of the game. On day 3 virtual ovens and fridges burst into flames. On day 4 a Sim refused to look for a job, and on day 5 the very same Sim found his first job. On day 6 many Sims already got promoted and fell in love, and on day 25 the first Sim became member of the Bundestag (the German parliament).

Each day was marked by specific events and the overall experience unfolds as a succession of these events. In fact, a further analysis showed that not surprisingly the valence of the events significantly influenced the participant's affective state after playing (r=.51, p<.01). This impacted the overall evaluation of *The Sims* at that day (r=.61, p<.01), which in turn predicted how much time participants wanted to spend with the game on the following day (lag 1, r=.41, p<.05).

Imagine how wrong it would be to take any of these 28 days and to use it as the sole indicator of product quality, without taking the other days into account. Some may view this as an example of how futile it is to rely on experiential measures. Obviously, the same game cannot be good on one day and bad on the other, right?! But it can and it actually is. Thus, quality is not in the single episodes, but in the way experience develops over time. Paradoxically, the moment we understand quality as a dynamic concept, we can make stable statements. In everyday game review language these are descriptions such as "loses appeal soon," "slow starter, but compensates by a deep and twisted story," "easy to master," or "epic." Each of these descriptions is about length and trajectory of the experience, about development, time and change, rather than about understanding a product as fixed.

28 days Sims had another lesson for me in stock. Longitudinal, dynamic perspectives on product evaluation are costly in terms of time and effort. They also require a commitment and discipline from participants, which is rare. A solution to this is retrospective interviewing. We simply let people reconstruct their overall trajectory after they had some time with a product. Our own try outs of this technique were promising. Margeritta von Wilamowitz Moellendorff and I (von Wilamowitz Moellendorff et al., 2006; von Wilamowitz-Moellendorff et al., 2007) developed an interviewing technique called CORPUS (Change-Oriented analysis of the Relationship between Product and User). For particular attributes, such as beauty, we asked participants to rate a product they had substantial contact with. We then asked them about the day of purchase and whether beauty was different back then. If there was change, we further reconstructed the shape of change and tried to identify particular "change incidents" responsible. For mobile phones, for example, we found an increase in usability, but also a decrease in novelty and stimulation. This was different for productivity software (Microsoft Office), which also increased in usability, but also in utility (people discover new possibilities) and in stimulation (there is always something new to discover). Through interviewing, we were not only able to reconstruct how specific product perceptions developed over time but also to uncover genre specific trajectories.

The implications of these examples are twofold. Time and change matter. Products are not stable. They may not literally shift shapes (although wear such as scratch marks are actual physical changes), but the experience created through these products is dynamic. Consequently, an experiential perspective on interactive products must accommodate change, must talk in terms of trajectories, in terms of experience extended over time. The downside is the methods. Longitudinal research is costly and the analysis of longitudinal quantitative data, for example, is not part of the statistics 101. And methods, which attempt to support the retrospective reconstruction of experiences and episodes from memory are especially challenging.

But you know all this already, don't you?! Because throughout the book which preceded this closing note, Evan just did an excellent job laying out the potential benefits and problems of "user experience over time". Well done!

Essen, March 2012 Marc Hassenzahl

References

Al-Azzawi, A., Frohlich, D., Wilson, M.: Beauty constructs for mp3 players. CoDesign 3(1), 59–74 (2007)

Anderson, S., Conway, M.: Investigating the structure of autobiographical memories. Learning Memory 19(5), 1178–1196 (1993)

Bailey, R.: Performance vs. preference. In: Human Factors and Ergonomics Society Annual Meeting Proceedings, vol. 37, pp. 282–286. Human Factors and Ergonomics Society (1993)

Barendregt, W., Bekker, M.M., Bouwhuis, D.G., Baauw, E.: Identifying usability and fun problems in a computer game during first use and after some practice. International Journal of Human-Computer Studies 64(9), 830–846 (2006)

Barsalou, L.: The content and organization of autobiographical memories. In: Neisser, U., Winograd, E. (eds.) Remembering Reconsidered: Ecological and Traditional Approaches to the Study of Memory, ch. 8, pp. 193–243. Cambridge University Press (1988)

Bartlett, F.: Remembering, Cambridge, MA (1932)

Battarbee, K.: Defining co-experience. In: Proceedings of the 2003 International Conference on Designing Pleasurable Products and Interfaces, pp. 109–113. ACM, New York (2003)

Battarbee, K., Koskinen, I.: Co-experience: user experience as interaction. CoDesign 1(1), 5–18 (2005)

Bech-Larsen, T., Nielsen, N.: A comparison of five elicitation techniques for elicitation of attributes of low involvement products. Journal of Economic Psychology 20, 315–341 (1999)

Bednarik, R., Myller, N., Sutinen, E., Tukiainen, M.: Effects of experience on gaze behaviour during program animation. In: Proceedings of the 17th Annual Psychology of Programming Interest Group Workshop (PPIG 2005), pp. 49–61 (2005)

Belk, R.W.: Possessions and the extended self. Journal of Consumer Research 15(2), 139–168 (1988)

Ben-Bassat, T., Meyer, J., Tractinsky, N.: Economic and subjective measures of the perceived value of aesthetics and usability. ACM Transactions on Computer-Human Interaction (TOCHI) 13(2), 234 (2006)

Berry, M., Drmac, Z., Jessup, E.: Matrices, vector spaces, and information retrieval. SIAM Review 41(2), 335–362 (1999)

Betsch, T., Plessner, H., Schwieren, C., Gutig, R.: I like it but I don't know why: A value-account approach to implicit attitude formation. Personality and Social Psychology Bulletin 27(2), 242 (2001)

Blei, D., Ng, A., Jordan, M.: Latent dirichlet allocation. The Journal of Machine Learning Research 3, 993–1022 (2003)

Blythe, M., Hassenzahl, M., Law, E., Vermeeren, A.: An analysis framework for user experience (UX) studies: A green paper. In: Law, E., Vermeeren, A., Hassenzahl, M., Blythe, M. (eds.) Towards a UX Manifesto-Proceedings of a Cost294-Affiliated Workshop on BHCI, pp. 1–5 (2007)

Blythe, M., Monk, A., Park, J.: Technology biographies: field study techinques for home use product development. In: Conference on Human Factors in Computing Systems, pp. 658–659. ACM, New York (2002)

Blythe, M., Overbeeke, K., Monk, A., Wright, P.: Funology: from usability to enjoyment. Kluwer Academic Publishers (2003)

Blythe, M., Reid, J., Wright, P., Geelhoed, E.: Interdisciplinary criticism: analysing the experience of riot! a location-sensitive digital narrative. Behaviour & Information Technology 25(2), 127–139 (2006)

Bolger, N., Davis, A., Rafaeli, E.: Diary methods: Capturing life as it is lived. Annual Review of Psychology 54(1), 579–616 (2003)

Bondarenko, O., Janssen, R.: Connecting visual cues to semantic judgments in the context of the office environment. Journal of the American Society for Information Science and Technology 60(5), 933–952 (2009)

Breivik, E., Supphellen, M.: Elicitation of product attributes in an evaluation context: A comparison of three elicitation techniques. Journal of Economic Psychology 24(1), 77–98 (2003)

Brombacher, A., Sander, P., Sonnemans, P., Rouvroye, J.: Managing product reliability in business processes under pressure. Reliability Engineering & System Safety 88(2), 137–146 (2005)

Burnham, K., Anderson, D.: Multimodel inference: understanding AIC and BIC in model selection. Sociological Methods & Research 33(2), 261 (2004)

Buxton, W.: Sketching user experiences: getting the design right and the right design. Morgan Kaufmann (2007)

Clarke, K.: Non-parametric multivariate analyses of changes in community structure. Australian Journal of Ecology 18(1), 117–143 (1993)

Cohen, J.: A power primer. Psychological Bulletin 112(1), 155 (1992)

Cohen, T., Widdows, D.: Empirical distributional semantics: Methods and biomedical applications. Journal of Biomedical Informatics 42(2), 390–405 (2009)

Collins, A., Loftus, E.: A spreading-activation theory of semantic processing. Psychological Review 82(6), 407–428 (1975)

Conway, M., Pleydell-Pearce, C.: The construction of autobiographical memories in the self-memory system. Psychological Review 107(2), 261–288 (2000)

Cooper, A.: The Inmates Are Running the Asylum. Macmillan Publishing Co., Inc. (1999)

Courage, C., Jain, J., Rosenbaum, S.: Best practices in longitudinal research. In: Proceedings of the 27th International Conference Extended Abstracts on Human Factors in Computing Systems, pp. 4791–4794. ACM, New York (2009)

Csikszentmihalyi, M., Rochberg-Halton, E.: The meaning of things: Domestic symbols and the self. Cambridge University Press (1981)

Davidoff, S., Lee, M.K., Yiu, C., Zimmerman, J., Dey, A.K.: Principles of Smart Home Control. In: Dourish, P., Friday, A. (eds.) UbiComp 2006. LNCS, vol. 4206, pp. 19–34. Springer, Heidelberg (2006)

Davis, F.D., Bagozzi, R.P., Warshaw, P.R.: User acceptance of computer technology: A comparison of two theoretical models. Management Science 35(8), 982–1003 (1989)

Davis, S., Carini, C.: Constructing a Player-Centred Definition of Fun for Video Games Design. In: HCI 2004 Conference, pp. 117–132. Springer (2004)
de Ridder, H.: Current issues and new techniques in visual quality assessment, pp. 869–872 (1996)
Deerwester, S., Dumais, S., Furnas, G., Landauer, T., Harshman, R.: Indexing by latent semantic analysis. Journal of the American Society for Information Science 41(6), 391–407 (1990)
Den Ouden, E.: Development of a design analysis model for consumer complaints. Ph.D. thesis, Eindhoven University of Technology (2006)
Den Ouden, E., Yuan, L., Sonnemans, P.J.M., Brombacher, A.C.: Quality and reliability problems from a consumer's perspective: an increasing problem overlooked by businesses? Quality and Reliability Engineering International 22(7), 821–838 (2006)
Desmet, P.: Designing emotions. Ph.D. thesis, Delft University of Technology (2002)
Desmet, P., Hekkert, P.: Framework of product experience. International Journal of Design 1(1), 57–66 (2007)
Desmet, P., Hekkert, P., Hillen, M.: Values and emotions; an empirical investigation in the relationship between emotional responses to products and human values (2004)
Dix, A., Finlay, J., Abowd, G., Beale, R.: Human-computer interaction. Prentice Hall (2004)
Draper, N., Smith, H.: Applied Regression Analysis. John Wiley and Sons Inc. (1998)
Du Gay, P., Hall, S., Janes, L.: Doing cultural studies: The story of the Sony Walkman. Sage Publications (1997)
Eckart, C., Young, G.: The approximation of one matrix by another of lower rank. Psychometrika 1(3), 211–218 (1936)
Edvardsson, B., Roos, I.: Critical incident techniques. International Journal of Service Industry Management 12(3), 251–268 (2001)
Egger, F.: From interactions to transactions: designing the trust experience for business-to-consumer electronic commerce. Ph.D. thesis, Eindhoven University of Technology (2003)
Erickson, T.: The design and long-term use of a personal electronic notebook: a reflective analysis. In: Proceedings of the SIGCHI Conference on Human Factors in Computing Systems: Common Ground, pp. 11–18 (1996)
Fallman, D., Waterworth, J.: Dealing with user experience and affective evaluation in HCI design: a repertory grid approach, pp. 2–7 (2005)
Fenko, A., Schifferstein, H., Hekkert, P.: Shifts in sensory dominance between various stages of user-product interactions. Applied Ergonomics 41(1), 34–40 (2009)
Festinger, L.: A Theory of Cognitive Dissonance. Stanford University Press (1957)
Flanagan, J.: The critical incident technique. Psychological Bulletin 51(4), 327–358 (1954)
Fleiss, J.L., Levin, B., Paik, M.C.: Statistical methods for rates and proportions. Wiley-Interscience (2003)
Forlizzi, J.: The product ecology: Understanding social product use and supporting design culture. International Journal of Design 2(1), 11–20 (2007)
Forlizzi, J., Battarbee, K.: Understanding experience in interactive systems. In: Proceedings of the 2004 Conference on Designing Interactive Systems: Processes, Practices, Methods, and Techniques, pp. 261–268 (2004)
Forlizzi, J., Ford, S.: The building blocks of experience: an early framework for interaction designers. In: Proceedings of the Conference on Designing Interactive Systems: Processes, Practices, Methods, and Techniques, pp. 419–423 (2000)
Forlizzi, J., Zimmerman, J., Evenson, S.: Crafting a Place for Interaction Design Research in HCI. Design Issues 24(3), 19–29 (2008)

Fox, C.: A stop list for general text. In: SIGIR Forum., vol. 24, pp. 19–21. ACM, New York (1989)

Fransella, F., Bell, R., Bannister, D.: A Manual for Repertory Grid Technique. Wiley (2003)

Frens, J.: Designing for rich interaction: Integrating form, interaction, and function. Ph.D. thesis, Eindhoven University of Technology (2006)

Frøkjær, E., Hertzum, M., Hornbæk, K.: Measuring usability: are effectiveness, efficiency, and satisfaction really correlated? In: Proceedings of the SIGCHI Conference on Human Factors in Computing Systems. ACM Press, The Hague (2000)

Funk, M., Rozinat, A., Karapanos, E., Karla Alves de Medeiros, A., Koca, A.: In situ evaluation of recommender systems: Framework and instrumentation. International Journal of Human-Computer Studies (2010) (to appear)

Funk, M., van der Putten, P.H.A., Corporaal, H.: Specification for user modeling with self-observing systems. In: Proceedings of the First International Conference on Advances in Computer-Human Interaction, pp. 243–248 (February 2008)

Furlan, R., Corradetti, R.: Reducing conjoint analysis paired comparisons tasks by a random selection procedure. Quality and Reliability Engineering International 22(5), 603–612 (2006)

Gaver, B., Dunne, T., Pacenti, E.: Cultural probes. Interactions 6(1), 21–29 (1999)

Gefen, D., Straub, D.: The relative importance of perceived ease of use in IS adoption: a study of e-commerce adoption. Journal of the Association for Information Systems 1(8), 1–28 (2000)

Gerken, J., Bak, P., Reiterer, H.: Longitudinal evaluation methods in human-computer studies and visual analytics. In: InfoVis 2007 Workshop on Metrics for the Evaluation of Visual Analytics (2007)

Glaser, B.G., Strauss, A.L.: The Discovery of Grounded Theory: Strategies for Qualitative Research. Aldine Transaction (1967)

Goldschmidt, G.: The dialectics of sketching. Creativity Research Journal 4(2), 123–143 (1991)

Green, P.E., Carmone Jr., F.J., Smith, S.M.: Multidimensional Scaling, Concepts and Applications. Allyn & Bacon (1989)

Groves, R., Fowler, F., Couper, M., Lepkowski, J., Singer, E., Tourangeau, R.: Survey methodology. John Wiley & Sons Inc. (2009)

Gutman, J.: A means-end chain model based on consumer categorization processes. The Journal of Marketing, 60–72 (1982)

Hallnäs, L., Redström, J.: From use to presence: on the expressions and aesthetics of everyday computational things. ACM Transactions on Computer-Human Interaction (TOCHI) 9(2), 106–124 (2002)

Hartmann, J., Sutcliffe, A., Angeli, A.D.: Towards a theory of user judgment of aesthetics and user interface quality. ACM Transactions on Computer-Human Interaction (TOCHI) 15(4), 1–30 (2008)

Hassenzahl, M.: The interplay of beauty, goodness, and usability in interactive products. Human-Computer Interaction 19(4), 319–349 (2004)

Hassenzahl, M.: The thing and i: understanding the relationship between user and product. Funology: from usability to enjoyment table of contents, 31–42 (2005)

Hassenzahl, M.: Hedonic, emotional, and experiential perspectives on product quality. In: Ghaoui, C. (ed.) Encyclopedia of Human Computer Interaction, pp. 266–272. Idea Group (2006)

Hassenzahl, M.: User experience (ux): towards an experiential perspective on product quality. In: IHM 2008: Proceedings of the 20th International Conference of the Association Francophone d'Interaction Homme-Machine, pp. 11–15. ACM, New York (2008)
Hassenzahl, M.: Experience Design: Technology for all the right reasons. Morgan and Claypool Publishers (2010)
Hassenzahl, M., Lai-Chong Law, E., Hvannberg, E.: User Experience-Towards a unified view. In: UX WS NordiCHI, vol. 6, pp. 1–3 (2006)
Hassenzahl, M., Platz, A., Burmester, M., Lehner, K.: Hedonic and ergonomic quality aspects determine a software's appeal. In: CHI 2000: Proceedings of the SIGCHI Conference on Human Factors in Computing Systems, pp. 201–208. ACM, New York (2000)
Hassenzahl, M., Schöbel, M., Trautmann, T.: How motivational orientation influences the evaluation and choice of hedonic and pragmatic interactive products: The role of regulatory focus. Interacting with Computers 20(4-5), 473–479 (2008)
Hassenzahl, M., Tractinsky, N.: User experience - a research agenda. Behaviour & Information Technology 25(2), 91–97 (2006)
Hassenzahl, M., Trautmann, T.: Analysis of web sites with the repertory grid technique (2001)
Hassenzahl, M., Ullrich, D.: To do or not to do: Differences in user experience and retrospective judgements depending on the presence or absence of instrumental goals. Interacting with Computers 19, 429–437 (2007)
Hassenzahl, M., Wessler, R.: Capturing design space from a user perspective: The repertory grid technique revisited. International Journal of Human-Computer Interaction 12(3), 441–459 (2000)
Haynes, S., Richard, D., Kubany, E.: Content validity in psychological assessment: A functional approach to concepts and methods. Psychological Assessment 7(3), 238–247 (1995)
Heidecker, S., Hassenzahl, M.: Eine gruppenspezifische repertory grid analyse der wahrgenommenen attraktivität von universitätswebsites. In: Gross, T. (ed.) Mensch & Computer 2007: Konferenz für interative und Kooperative Medien, pp. 129–138 (2007)
Hekkert, P., Leder, H.: Product aesthetics. In: Schifferstein, H., Hekkert, P. (eds.) Product Experience, pp. 259–286 (2008)
Hektner, J.M., Schmidt, J.A., Csikszentmihalyi, M.: Experience Sampling Method: Measuring the Quality of Everyday Life. Sage Publications Inc. (2007)
Hertzum, M., Clemmensen, T., Hornbæk, K., Kumar, J., Shi, Q., Yammiyavar, P.: Usability Constructs: A Cross-Cultural Study of How Users and Developers Experience Their Use of Information Systems. In: Aykin, N. (ed.) HCII 2007. LNCS, vol. 4559, pp. 317–326. Springer, Heidelberg (2007)
Hinkin, T.: A review of scale development practices in the study of organizations. Journal of Management 21(5), 967 (1995)
Hofstede, G.: Culture's consequences: Comparing values, behaviors, institutions, and organizations across nations. Sage Publications (2001)
Hornbæk, K.: Usability Evaluation as Idea Generation. In: Law, E.L.-C., Hvannberg, E.T., Cockton, G. (eds.) Maturing Usability, pp. 267–286. Springer (2008)
Hornbæk, K., Law, E.: Meta-analysis of correlations among usability measures. In: Proceedings of the SIGCHI Conference on Human Factors in Computing Systems, pp. 617–626. ACM, New York (2007)
Hsieh, G., Li, I., Dey, A., Forlizzi, J., Hudson, S.: Using visualizations to increase compliance in experience sampling. In: Proceedings of the 10th International Conference on Ubiquitous Computing, pp. 164–167. ACM (2008)
Hsieh, H.F., Shannon, S.E.: Three approaches to qualitative content analysis. Qualitative Health Research 15(9), 1277–1288 (2005)

IJsselsteijn, W., van Baren, J., Markopoulos, P., Romero, N., de Ruyter, B.: Measuring affective benefits and costs of mediated awareness: Development and validation of the abc-questionnaire. In: Markopoulos, P., de Ruyter, B., Mackay, W. (eds.) Awareness Systems: Advances in Theory, Methodology and Design, pp. 473–488. Springer (2009)

ISO, ISO 9241 - Ergonomic requirements for office work with visual display terminals (VDTs) - Part 11: Guidance on usability. International Organization for Standardization (1996)

Janlert, L.E., Stolterman, E.: The character of things. Design Studies 18(3), 297–314 (1997)

Jordan, P.: Designing Pleasurable Products: An Introduction to New Human Factors. Taylor & Francis (2000)

Jordan, P., Persson, S.: Exploring users' product constructs: how people think about different types of product. International Journal of CoCreation in Design and the Arts 3(1), 97–106 (2007)

Kahneman, D.: Objective happiness. In: Kahneman, D., Diener, E., Schwarz, N. (eds.) Well-Being: The Foundations of Hedonic Psychology, pp. 3–25. Russel Sage Foundation, New York (1999)

Kahneman, D., Krueger, A.B., Schkade, D.A., Schwarz, N., Stone, A.A.: A survey method for characterizing daily life experience: The day reconstruction method 306(5702), 1776–1780 (2004)

Kahneman, D., Slovic, P., Tversky, A.: Judgment under uncertainty: Heuristics and biases. Cambridge University Press (1982)

Karapanos, E., Hassenzahl, M., Martens, J.-B.: User experience over time. In: CHI 2008 Extended Abstracts on Human Factors in Computing Systems, pp. 3561–3566. ACM, Florence (2008a)

Karapanos, E., Martens, J.-B.: Characterizing the Diversity in Users' Perceptions. In: Baranauskas, C., Abascal, J., Barbosa, S.D.J. (eds.) INTERACT 2007. LNCS, vol. 4662, pp. 515–518. Springer, Heidelberg (2007)

Karapanos, E., Martens, J.-B.: The quantitative side of the repertory grid technique: some concerns. In: Proceedings of the Workshop Now Let's Do It in Practice: User Experience Evaluation Methods in Product Development, Human Factors in Computing Systems, CHI 2008, Florence (2008)

Karapanos, E., Martens, J.-B., Hassenzahl, M.: Accounting for diversity in subjective judgments. In: CHI 2009: Proceedings of the 27th International Conference on Human Factors in Computing Systems, pp. 639–648. ACM, New York (2009a)

Karapanos, E., Martens, J.-B., Hassenzahl, M.: Reconstructing Experiences through Sketching (2009b), arXiv pre-print available at `http://arxiv.org/abs/0912.5343`

Karapanos, E., Wensveen, S., Friederichs, B., Martens, J.-B.: Do knobs have character?: exploring diversity in users' inferences. In: CHI 2008 Extended Abstracts on Human Factors in Computing Systems. ACM, Florence (2008b)

Karapanos, E., Zimmerman, J., Forlizzi, J., Martens, J.-B.: User experience over time: an initial framework. In: CHI 2009: Proceedings of the 27th International Conference on Human Factors in Computing Systems, pp. 729–738. ACM, New York (2009c)

Katsanos, C., Tselios, N., Avouris, N.: Automated semantic elaboration of web site information architecture. Interacting with Computers 20(6), 535–544 (2008)

Kaur, I., Hornof, A.: A comparison of LSA, wordNet and PMI-IR for predicting user click behavior. In: Proceedings of the SIGCHI Conference on Human Factors in Computing Systems, pp. 51–60. ACM, New York (2005)

Kelly, G.A.: The Psychology of Personal Constructs. Norton, New York (1955)

Kelly, G.A.: A mathematical approach to psychology. In: Maher, B. (ed.) Clinical Psychology and Personality: The Selected Papers of George Kelly, pp. 94–113. Wiley, New York (1969)

Khan, V., Markopoulos, P., Eggen, B., Ijsselsteijn, W., de Ruyter, B.: Reconexp: a way to reduce the data loss of the experiencing sampling method. In: Proceedings of the 10th International Conference on Human Computer Interaction with Mobile Devices and Services, pp. 471–476. ACM (2008)

Kim, S., Malhotra, N.: A longitudinal model of continued IS use: An integrative view of four mechanisms underlying postadoption phenomena. Management Science 51(5), 741–755 (2005)

King, G., Keohane, R., Verba, S.: Designing social inquiry: Scientific inference in qualitative research. Princeton University Press, Princeton (1994)

Kissel, G.: The effect of computer experience on subjective and objective software usability measures. In: Conference on Human Factors in Computing Systems, pp. 284–285. ACM, New York (1995)

Kjeldskov, J., Skov, M., Stage, J.: A longitudinal study of usability in health care: Does time heal? International Journal of Medical Informatics (2008) (in press)

Koca, A., Funk, M.: Technical and Commercial Feasibility Study for UXSUITE, A Market Research Software Toolset for Preventing No Fault Found. Application number 10926 (2009)

Koca, A., Funk, M., Karapanos, E., Rozinat, A., van der Aalst, W., Corporaal, H., Martens, J., van der Putten, P., Weijters, A., Brombacher, A.: Soft reliability: an interdisciplinary approach with a user-system focus. Quality and Reliability Engineering International 25(1), 3–20 (2009)

Kontostathis, A.: Essential dimensions of latent semantic indexing (lsi). In: 40th Annual Hawaii International Conference on System Sciences, HICSS 2007, p. 73 (2007)

Koriat, A., Goldsmith, M., Pansky, A.: Toward a psychology of memory accuracy. Annual Review of Psychology 51(1), 481–537 (2000)

Krippendorff, K.: Content Analysis: An Introduction to Its Methodology. Sage Publications (2004)

Kujala, S., Kauppinen, M.: Identifying and selecting users for user-centered design. In: Proceedings of the Third Nordic Conference on Human-Computer Interaction. ACM, Tampere (2004)

Kujala, S., Roto, V., Vnnen-Vainio-Mattila, K., Karapanos, E., Sinnel, A.: Ux curve: A method for evaluating long-term user experience. Interacting with Computers 23(5), 473–483 (2011); Feminism and HCI: New Perspectives, http://www.sciencedirect.com/science/article/pii/S0953543811000737

Kurosu, M., Kashimura, K.: Apparent usability vs. inherent usability: experimental analysis on the determinants of the apparent usability. In: Conference on Human Factors in Computing Systems, pp. 292–293. ACM, New York (1995)

Landauer, T., Dumais, S.: A solution to Plato's problem: The latent semantic analysis theory of acquisition, induction, and representation of knowledge. Psychological Review 104(2), 211–240 (1997)

Landauer, T., Laham, D., Foltz, P.: Automated scoring and annotation of essays with the Intelligent Essay Assessor, pp. 87–112. Lawrence Erlbaum Associates, Inc., Mahwah (2003)

Larsen, K., Monarchi, D., Hovorka, D., Bailey, C.: Analyzing unstructured text data: Using latent categorization to identify intellectual communities in information systems. Decision Support Systems 45(4), 884–896 (2008a)

Larsen, K., Nevo, D., Rich, E.: Exploring the Semantic Validity of Questionnaire Scales. In: Proceedings of the 41st Annual Hawaii International Conference on System Sciences, p. 440 (2008b)
Lavie, T., Tractinsky, N.: Assessing dimensions of perceived visual aesthetics of web sites. International Journal of Human-Computer Studies 60(3), 269–298 (2004)
Lee, Y., Kozar, K., Larsen, K.: The Technology Acceptance Model: Past, Present, and Future. Communications of the Association for Information Systems 12(1), 752–780 (2003)
Light, A.: Adding method to meaning: a technique for exploring peoples' experience with technology. Behaviour & Information Technology 25(2), 175–187 (2006)
Likert, R.: A technique for the measurement of attitudes. Archives of Psychology, vol. 140 (1932)
Lindgaard, G., Dudek, C.: What is this evasive beast we call user satisfaction? Interacting with Computers 15, 429–452 (2003)
Lindgaard, G., Fernandes, G., Dudek, C., Brown, J.: Attention web designers: You have 50 milliseconds to make a good first impression! Behaviour & Information Technology 25(2), 115–126 (2006)
Lyons, J.: Semantics. Cambridge University Press (1977)
Machotka, P.: Daniel berlyne's contributions to empirical aesthetics. Motivation and Emotion 4(2), 113–121 (1980)
Mahlke, S.: Aesthetic and symbolic qualities as antecedents of overall judgements of interactive products. In: People and Computers XX - Engage, Proceedings of HCI 2006, pp. 57–64. Springer, London (2006)
Mahlke, S., Lindgaard, G.: Emotional Experiences and Quality Perceptions of Interactive Products. In: Jacko, J.A. (ed.) HCI 2007, Part I. LNCS, vol. 4550, pp. 164–173. Springer, Heidelberg (2007)
Mahlke, S., Thüring, M.: Studying antecedents of emotional experiences in interactive contexts. In: Proceedings of the SIGCHI Conference on Human Factors in Computing Systems, pp. 915–918 (2007)
Markopoulos, P., Romero, N., van Baren, J., Ijsselsteijn, W., de Ruyter, B., Farshchian, B.: Keeping in touch with the family: home and away with the astra awareness system, pp. 1351–1354. ACM Press, New York (2004)
Martens, J.-B.: Image technology design: A perceptual approach. Kluwer Academic Publisher (2003)
Martens, J.-B.: An algorithmic procedure for the analysis of Repertory Grid Data. Personal Communication (October 2009a)
Martens, J.-B.: A tentative framework of the design process. Personal Communication (October 2009b)
Maslow, A.: A theory of human motivation. Psychological Review 50, 370–396 (1946)
McCarthy, J., Wright, P.: Technology As Experience. MIT Press (2004)
Means, B., Nigam, A., Zarrow, M., Loftus, E., Donaldson, M.: Autobiographical memory for health-related events. Vital Health Stat. 6(2), 1–22 (1989)
Mendoza, V., Novick, D.G.: Usability over time. In: Proceedings of the 23rd Annual International Conference on Design of Communication: Documenting & Designing for Pervasive Information, pp. 151–158 (2005)
Minge, M.: Dynamics of User Experience. In: Proceedings of the Workshop on Research Goals and Strategies for Studying User Experience and Emotion, NordiCHI 2008 (2008)
Neisser, U.: John Dean's memory: A case study. Cognition 9(1), 1–22 (1981)
Nielsen, J., Bellcore, M.: The usability engineering life cycle. Computer 25(3), 12–22 (1992)

Nielsen, J., Desurvire, H.: Comparative design review: an exercise in parallel design. In: Proceedings of the INTERACT 1993 and CHI 1993 Conference on Human Factors in Computing Systems, pp. 414–417. ACM, New York (1993)

Nielsen, J., Levy, J.: Measuring usability: preference vs. performance. Commun. ACM 37(4), 66–75 (1994)

Norman, D.: The psychology of everyday things. Basic books, New York (1988)

Norman, D.: Emotional design: Why we love (or hate) everyday things. Basic Books, New York (2004)

Norman, D.: The Way I See it. Memory is more important than actuality. Interactions 16(2), 24–26 (2009)

Oliver, R.L.: A cognitive model of the antecedents and consequences of satisfaction decisions. Journal of Marketing Research 17(4), 460–469 (1980)

Osgood, C., Suci, G., Tannenbaum, P.: The measurement of meaning. University of Illinois Press (1957)

Porter, M.F.: An algorithm for suffix stripping 14, 130–137 (1980)

Prümper, J., Zapf, D., Brodbeck, F., Frese, M.: Some surprising differences between novice and expert errors in computerized office work. Behaviour & Information Technology 11(6), 319–328 (1992)

Rafaeli, A., Vilnai-Yavetz, I.: Instrumentality, aesthetics and symbolism of physical artifacts as triggers of emotion. Theoretical Issues in Ergonomics Science 5(1), 91–112 (2004)

Ramsay, J.: Matfit: A fortran subroutine for comparing two matrices in a subspace. Psychometrika 55(3), 551–553 (1990)

Redelmeier, D.A., Kahneman, D.: Patients' memories of painful medical treatments: real-time and retrospective evaluations of two minimally invasive procedures. Pain 66(1), 3–8 (1996),
http://www.sciencedirect.com/science/article/B6T0K-3R7BC4G-K/2/0f93571518134b4616b3d32da94f0dca

Reyna, V., Kiernan, B.: Development of gist versus verbatim memory in sentence recognition: Effects of lexical familiarity, semantic content, encoding instructions, and retention interval. Developmental Psychology 30(2), 178 (1994)

Reynolds, T., Gutman, J.: Laddering Theory, Method, Analysis and Interpretation. Journal of Advertising Research 28(1), 11–31 (1988)

Robinson, M., Clore, G.: Belief and feeling: Evidence for an accessibility model of emotional self-report. Psychological Bulletin 128(6), 934–960 (2002)

Russell, J.: A circumplex model of affect. Journal of Personality and Social Psychology 39(6), 1161–1178 (1980)

Salton, G., Buckley, C.: Term-weighting approaches in automatic text retrieval. Information Processing & Management 24(5), 513–523 (1988)

Salton, G., Wong, A., Yang, C.: A vector space model for automatic indexing. Communications of the ACM 18(11), 613–620 (1975)

Sandt Van de, U.: Incomplete paired comparisons using balanced lattice designs. Journal of Marketing Research 7, 246–248 (1970)

Sanford, A., Sanford, A., Molle, J., Emmott, C.: Shallow processing and attention capture in written and spoken discourse. Discourse Processes 42(2), 109–130 (2006)

Scandura, T., Williams, E.: Research methodology in management: Current practices, trends, and implications for future research. Academy of Management Journal 43(6), 1248–1264 (2000)

Schenkman, B.N., Jönsson, F.U.: Aesthetics and preferences of web pages. Behaviour & Information Technology 19, 367–377 (2000)

Schrepp, M., Held, T., Laugwitz, B.: The influence of hedonic quality on the attractiveness of user interfaces of business management software. Interacting with Computers 18(5), 1055–1069 (2006)

Schwartz, S.: Universals in the content and structure of values: theoritical advances and empirical tests in 20 countries. In: Zanna, M. (ed.) Advances in Experimental Psychology, vol. 25, pp. 1–65. Academic Press (1992)

Schwarz, N., Kahneman, D., Xu, J., Belli, R., Stafford, F., Alwin, D.: Global and episodic reports of hedonic experience. Calendar and Time Diary Methods in Life Course Research: Methods in Life Course Research 157 (2008)

Silverstone, R., Haddon, L.: Design and the domestication of information and communication technologies: Technical change and everyday life. In: Silverstone, R., Mansell, R. (eds.) Communication by Design: The Politics of Information and Communication Technologies, pp. 44–74. Oxford University Press (1996)

Sonnemans, J., Frijda, N.H.: The structure of subjective emotional intensity. Cognition & Emotion 8(4), 329–350 (1994), http://www.tandfonline.com/doi/abs/10.1080/02699939408408945

Steenkamp, J.-B., Van Trijp, H.: Attribute elicitation in marketing research: A comparison of three procedures. Marketing Letters 8(2), 153–165 (1997)

Strauss, A.L., Corbin, J.M.: Basics of Qualitative Research: Techniques and Procedures for Developing Grounded Theory. Sage Publications Inc. (1998)

Suchman, L.: Human-machine reconfigurations: Plans and situated actions. Cambridge University Press (2006)

Suri, J.: Designing Experience: Whether to Measure Pleasure. In: Green, W., Jordan, P. (eds.) Pleasure with Products: Beyond Usability, pp. 161–174. Taylor & Francis, London (2002)

Szostek, A., Karapanos, E., Eggen, B., Holenderski, M.: Understanding the implications of social translucence for systems supporting communication at work. In: Proceedings of the ACM 2008 Conference on Computer Supported Cooperative Work, pp. 649–658. ACM, New York (2008)

Taylor, A.S., Swan, L.: Artful systems in the home (2005)

Tractinsky, N.: Aesthetics and apparent usability: Empirically assessing cultural and methodological issues. In: Conference on Human Factors in Computing Systems, pp. 115–122. ACM, New York (1997)

Tractinsky, N., Cokhavi, A., Kirschenbaum, M., Sharfi, T.: Evaluating the consistency of immediate aesthetic perceptions of web pages. International Journal of Human-Computer Studies 64(11), 1071–1083 (2006)

Tractinsky, N., Katz, A.S., Ikar, D.: What is beautiful is usable. Interacting with Computers 13(2), 127–145 (2000)

Tractinsky, N., Zmiri, D.: Exploring attributes of skins as potential antecedents of emotion in hci. In: Fishwick, P. (ed.) Aesthetic Computing. MIT Press (2006)

Tulving, E.: EPISODIC MEMORY: From Mind to Brain. Annual Review of Psychology 53(1), 1–25 (2002)

Väänänen-Vainio-Mattila, K., Roto, V., Hassenzahl, M.: Now let's do it in practice: user experience evaluation methods in product development. In: CHI 2008: CHI 2008 Extended Abstracts on Human Factors in Computing Systems, pp. 3961–3964. ACM, New York (2008)

Van Bel, D.T., Ijsselsteijn, W.A., de Kort, Y.A.: Interpersonal connectedness: conceptualization and directions for a measurement instrument. In: CHI 2008: CHI 2008 Extended Abstracts on Human Factors in Computing Systems, pp. 3129–3134. ACM, New York (2008) van de Garde-Perik, E.: Ambient intelligence & personalization: people's perspectives on information privacy. Ph.D. thesis. Eindhoven University of Technology (2008)

van der Aalst, W.M.P., van Dongen, B.F., Günther, C.W., Mans, R.S., de Medeiros, A.K.A., Rozinat, A., Rubin, V., Song, M., Verbeek, H.M.W(E.), Weijters, A.J.M.M.T.: ProM 4.0: Comprehensive Support for *Real* Process Analysis. In: Kleijn, J., Yakovlev, A. (eds.) ICATPN 2007. LNCS, vol. 4546, pp. 484–494. Springer, Heidelberg (2007)

van Kleef, E., van Trijp, H., Luning, P.: Consumer research in the early stages of new product development: a critical review of methods and techniques. Food Quality and Preference 16(3), 181–202 (2005)

Van Schaik, P., Ling, J.: Design parameters of rating scales for web sites. ACM Transactions on Computer-Human Interaction (TOCHI) 14(1), 4 (2007)

Van Schaik, P., Ling, J.: Modelling user experience with web sites: Usability, hedonic value, beauty and goodness. Interacting with Computers 20(3), 419–432 (2008)

Vaughan, M., Courage, C., Rosenbaum, S., Jain, J., Hammontree, M., Beale, R., Welsh, D.: Longitudinal usability data collection: art versus science? (2008)

Venkatesh, V., Davis, F.: A theoretical extension of the technology acceptance model: four longitudinal field studies. Management Science 46(2), 186–204 (2000)

Venkatesh, V., Johnson, P.: Telecommuting technology implementations: A within-and between-subjects longitudinal field study. Personnel Psychology 55(3), 661–687 (2002)

Venkatesh, V., Morris, M.G., Davis, G.B., Davis, F.D.: User acceptance of information technology: Toward a unified view. MIS Quarterly 27(3), 425–478 (2003)

von Wilamowitz Moellendorff, M., Hassenzahl, M., Platz, A.: Dynamics of user experience: How the perceived quality of mobile phones changes over time. In: User Experience - Towards a Unified View, Workshop at the 4th Nordic Conference on Human-Computer Interaction, pp. 74–78 (2006)

von Wilamowitz-Moellendorff, M., Hassenzahl, M., Platz, A.: In: Gross, T. (ed.) Mensch & Computer, pp. 49–58 (2007)

Wright, P., Blythe, M.: User experience research as an inter-discipline: Towards a UX Manifesto. In: Law, E., Vermeeren, A., Hassenzahl, M., Blythe, M. (eds.) Towards a UX Manifesto-Proceedings of a Cost294-Affiliated Workshop on BHCI, pp. 65–70 (2007)

Wright, P., McCarthy, J.: Technology as experience. The MIT Press (2004)

Wright, P., McCarthy, J.: Empathy and experience in hci. In: CHI 2008: Proceeding of the Twenty-Sixth Annual SIGCHI Conference on Human Factors in Computing Systems, pp. 637–646. ACM, New York (2008)

Yorke, M.: Bipolarity or not? some conceptual problems relating to bipolar rating scales. British Educational Research Journa 27(2), 171–186 (2001)

Zimmerman, J.: Designing for the self: making products that help people become the person they desire to be. In: CHI 2009: Proceedings of the 27th International Conference on Human Factors in Computing Systems, pp. 395–404. ACM, New York (2009)

Zimmerman, J., Forlizzi, J., Evenson, S.: Research through design as a method for interaction design research in HCI. In: Proceedings of the SIGCHI Conference on Human Factors in Computing Systems, pp. 493–502. ACM, New York (2007)

Zimmerman, J., Forlizzi, J., Koskinen, I.: Building a unified framework for the practice of experience Design. In: Proceedings of the 27th International Conference Extended Abstracts on Human Factors in Computing Systems, pp. 4803–4806. ACM, New York (2009)